Cosmological Implications of
Heisenberg's Principle

Cosmological Implications of
Heisenberg's Principle

Julio A Gonzalo
Universidad Autónoma de Madrid, Spain

World Scientific

NEW JERSEY · LONDON · SINGAPORE · BEIJING · SHANGHAI · HONG KONG · TAIPEI · CHENNAI

Published by

World Scientific Publishing Co. Pte. Ltd.

5 Toh Tuck Link, Singapore 596224

USA office: 27 Warren Street, Suite 401-402, Hackensack, NJ 07601

UK office: 57 Shelton Street, Covent Garden, London WC2H 9HE

Library of Congress Cataloging-in-Publication Data
Gonzalo, Julio A. (Julio Antonio), author.
   Cosmological implications of Heisenberg's principle / Julio A. Gonzalo,
Universidad Autónoma de Madrid, Spain.
      pages cm
   Includes bibliographical references and index.
   ISBN 978-9814675376 (hardcover : alk. paper) -- ISBN 9814675377 (hardcover : alk. paper)
   1. Heisenberg uncertainty principle. 2. Quantum theory. 3. Cosmology. I. Title.
   QC174.17.H4G66 2015
   530.12--dc23

                                                                    2015031774

British Library Cataloguing-in-Publication Data
A catalogue record for this book is available from the British Library.

In-house Editor: Christopher Teo

Typeset by Stallion Press
Email: enquiries@stallionpress.com

Printed in Singapore

PREFACE

The implications of Heisenberg's Uncertainty Principle for the study of the only universe amenable to physical observation (unlike the infinite unobservable "multiverses" so fashionable these days) are very substantial as I hope to show in this short book.

The book begins in Chapter 1 with an invited lecture on *"Planck, Einstein, and Mach"* "given at the Pontificium Ateneum *"Regina Apostolorum"* in a Symposium to commemorate the fifth anniversary of Professor Stanley L. Jaki death. In that lecture I tried to put in a proper perspective the monumental contributions of Max Planck and Albert Einstein to 20th century physics and to show how both of them, early admirers of Ernst Mach, became diametrically opposed to Mach's positivist, materialist world views.

In chapters 2 to 5, brief biographical sketches of the main protagonists of Quantum Mechanics (Bohr, Sommerfeld, De Broglie, Schrödinger, Heisenberg, Jordan, Dirac, Born, Debye, Compton, Pauli, Bose, Fermi and Wigner) are given. Half of them are written by S.L. Jaki and by myself, and the rest are taken from Nobel Lectures (Nobelprize.org) to whom credit must be given here again. They were collected in *"Pioneers of Quantum Physics"* (J.A. Gonzalo, S.L. Jaki, J.C. del Valle, eds) (Ciencia y Cultura: Madrid, 2011).

Chapter 6, on *"Indeterminacy vs Uncertainty"*, is devoted to show why Heisenberg's Principle is on *"uncertainty"* in knowledge, rather than on *"indeterminacy"* in reality itself.

Chapter 7 reviews briefly how Planck found $\hbar$ the quantum of action, and how he recognized as universal constants k_B

(Boltzmann's constant) and c (velocity of light), introducing in 20th century physics for the first time the all important concept of *universal constant*.

In Chapter 8 are introduced in parallel with *Planck's "natural" units* (based upon a few universal constants), what may be called *Heisenberg-Lemaitre* units, based upon Heisenberg's principle and $M_u \cong 10^{55}\,g$, the mass of the entire universe, with about 10^{11}- 10^{12} galaxies, each encompassing about 10^{11}- 10^{12} stars, each with an average mass of the order of the Sun mass, $M_s \cong 2 \cdot 10^{33}$.

Chapter 9 is devoted to discuss the decisive implications of a *finite universe*.

In Chapter 10 it is shown that the *zero-point energy* of electromagnetic radiation in the universe is enormous but finite, and it is shown also that, after decoupling (atom formation), the radiation energy density decreases as $\rho_\gamma(T) = \rho_\gamma(T_{dec})(T/T_C)$, while zero point energy density $\rho_{zp}(T) = \rho_{zp}(T_{dec})(1 - T/T_C)$, insuring therefore that total energy density is conserved through cosmic expansion.

In Chapter 11 recent work by Julio A. Gonzalo and Manuel Alfonseca (2014) entitled *"Comment of the 1% Concordance Hubble Constant"*, discusses quantitatively compact solutions of Einstein's cosmological equations for a flat, Lambda Cold Dark Matter – (LCDM) universe and for an open (Friedman-Lemaitre) universe, and shows that the latter can be reconciled in principle with the observational evidence and that it has definite advantages to justify the largest observable galactic redshifts $(z \cong 10.9)$.

Chapter 12 discusses *"dark matter"*, showing that it is not correct to expect Keplerian velocities for stars circulating around spiral galaxies, and the so called *"dark energy"* which could be attributed (in principle) to $k < 0$ (space-time curvature) as well as to $\Lambda > 0$ (non-zero cosmological constant) and, finally, cosmic *"accelerated expansion"*.

To conclude, Chapter 13 discusses the *philosophical meaning* of Heisenberg's uncertainty principle and its positive *usefulness* to gain qualitative and quantitative information on physical reality.

I would like to thank Fr. Stanley L. Jaki, OSB (r.i.p.), Fr. Manuel M. Carreira SJ, Professor Manuel Alfonseca, and Professor Antonio Alfonso-Fauss for many informative discussions

Madrid, 1 November 2014
Festivity of All Saints
JAG

CONTENTS

Preface v

Prologue 1

1. Planck, Einstein and Mach.. 5

2. About the origins of Quantum Mechanics I...................... 17

3. About the origins of Quantum Mechanics II 29

4. About the origins of Quantum Mechanics III................... 45

5. About the origins of Quantum Mechanics IV................... 64

6. Indeterminacy vs Uncertainty.................................... 78

7. The universal constants.. 84

8. Planck's units and Heisenberg-Lemaitre units 91

9. Implications of a finite universe 98

10. Cosmic zero-point energy...................................... 106

11. Rigorous solutions of Einstein's cosmological equation...... 111

12. On the evidence for dark matter, dark energy &
    accelerated expansion.. 121

13. On Physics and Philosophy.................................... 129

Appendix: From Scientific Cosmology to a Created
    Universe... 137

Chronology 154

Glossary 158

Authour Index.. 173

Subject Index 177

PROLOGUE

(Interview with Professor Manuel Alfonseca, U.A.M.)

You have been a professor of Computational Sciences at the U.A.M and the U.P.M for many years. You are also the author of several books of science popularization and have lectured extensively on the origin of the universe and related matters. How would you recapitulate the astrophysical evidence for the Big Bang theory?

Essentially, the Big Bang theory is based on two evidences, both foreseen by Ralph Alpher and his colleagues in 1948, and confirmed experimentally during the sixties:

- The average composition of the universe (about 75% hydrogen and 25% helium, with a little deuterium and traces of other elements). The Big Bang theory predicts that protons would fuse spontaneously into deuterium and helium (with a little lithium) a couple of minutes after the Big Bang, giving rise to the observed composition.

- The cosmic microwave background radiation, which Alpher estimated would have an average temperature of about 5K. Several hundred thousand years after the Big

Bang, when the temperature and pressure would have gone down sufficiently, protons captured electrons, making the universe transparent. The resulting energy flash has come to us as microwave radiation, as a result of the red shift due to the universal expansion. This radiation was discovered in the sixties at a little under 4K, almost identical to Alpher's prediction.

Surprisingly enough, Ralph Alpher was never awarded the Nobel Prize after his two predictions were confirmed.

Was there any matter or energy before the Big Bang?

We don't have an answer to that question. We don't even know whether time existed before the Big Bang, so the question may even be nonsensical. There are several proposed quantum gravity theories, some of which assert that time just at the beginning would have been imaginary, therefore behaving like space, but none of these theories has had any experimental confirmation.

Some scientists say that, even if present science cannot explain satisfactorily all the details of cosmic evolution, it will likely do it in the near future. What do you think?

I think that we will never know everything. Science advances, but every new discovery opens new questions and fields of research. Scientific research is a never-ending task.

Is it likely that the Big Bang cosmology will be overthrown by new scientific evidence in the near future?

As every scientific theory, the Big Bang cosmology can always be replaced by a better theory. However, when this happened to Newton's theory of gravitation, the new theory (general relativity) did not demolish Newton's gravitation, it just made it into a first approximation, which still can be applied for practical purposes. I think something similar may happen to the Big Bang cosmology, which is based on general relativity and two accurate confirmed predictions. However, we must take into account that general relativity will have to be replaced one day, as it cannot be the final theory (assuming that something like that exists). We have now two different incompatible physical theories: general relativity and quantum mechanics, and they can co-exist smoothly because they are usually applied to very different objects (the first to stars and galaxies, the other to elementary particles and atoms). However, when we get nearer to the Big Bang, both fields (microscopic and macroscopic) tend to mix, and therefore the two theories interfere with one another. Just now we don't have an answer to this problem. When we have (if ever), the Big Bang theory will have to be re-thought.

Some noted scientists say that the universe came out of nothing. What do you think about it?

That they confuse nothing with the vacuum. We know since 2500 years ago (Parmenides was the first to say it) that nothing cannot have any property, including existence. Therefore nothing does not exist. And from what does not exist, nothing can come. On the other hand, the vacuum (which is very different from nothing) does have properties: space, time, energy... Quantum mechanics and the Heisenberg principle predict that the vacuum contains a certain amount of energy. However, the Schwinger effect (the spontaneous creation of particle pairs out of the vacuum by applying an electric field), which was predicted starting from this consideration, has not yet been confirmed, not even for electron-positron pairs, which would be the easiest to generate. Therefore the assumption that a whole universe like ours could have come spontaneously out of the vacuum (never from nothing) seems a little far-fetched, given the current state of our knowledge.

As a scientist and a practicing Catholic, do you see any direct contradiction between what you know as a scientist and what your reason and your faith tell you about the cosmos, man and the Creator?

None at all. Science and religion address different questions with different methods. No one can prove using scientific methods God's existence or inexistence. Religion just tells us that God created the universe, not how. No scientific discovery can contradict that assertion, even if it were found that, after all, the universe did not have a beginning. In fact, Thomas Aquinas said, several centuries ago, that *creatio originans* (creation at a given instant of time) is undecidable by reason, therefore any scientific discovery in this direction would not affect my religious beliefs. The same happens with the multiverse theories, so in vogue. If God could create one universe, why couldn't He create two, or 10^{500}? Even biological evolution, which is usually presented as a problem for believers, has never been so for Catholics.

1. PLANCK, EINSTEIN AND MACH[1]

Summary

Stanley L. Jaki dedicates two full chapters[1] in his "The Road of Science and the Ways to God" (The University of Chicago Press, 1978) to show convincingly that the philosophical foundations of the two greatest physicists of the 20th century were realist rather than positivist or idealist. In chapter 11, "The Quantum of the Science", devoted to Max Planck, Fr. Jaki shows how the pioneer of Quantum Theory, who was at the beginning very respectful of Ernst Mach, the epitome of scientific relativism, ends up in the opposite camp, defending vigorously scientific objectivity and realism. In chapter 12, "The Quantity of the Universe" devoted to Albert Einstein, Fr. Jaki demonstrates also convincingly, that the father of the Theory of Relativity, also originally an admirer of Mach, ended up side by side with Planck, defending objective reality in the natural world, and contradicting the utterly relativistic approach of Mach, which reduced everything to sensory perceptions. In the opening decades of the 21th century, with numerous scientists defending purely positivistic and materialistic views of man and cosmos, it is fit to remind all of us that Fr. Jaki did a monumental service to the cause of scientific realism documenting unambiguously that the scientific views of the true pioneers of Modern Physics, Planck and Einstein, were and are

---

compatible with a natural world orderly and well done, and with an intellectual capability in man well suited to investigate and understand it. Modern science is therefore a monumental proof that the natural world as well as man's intellect, are contingent and are due to an all-powerful and intelligent Creator.

Introduction

It is no secret that Max Planck (1958-1947) and Albert Einstein (1979-1955) were the two greatest physicists of the 20th century.

In this work we will introduce the creative scientific contributions of Planck and Einstein before, following Fr. Jaki, going in more detail into the respective stories of how each one (Planck and Einstein), managed to get rid of the prevailing positivistic world view in Germany at the end of the 19th century, exemplified by Mach's positivism.

The deep and painstaking historical investigations of Fr. Jaki provide abundant evidence of the process by which Planck and Einstein, each in his own way, both driven by the clear evidence of their own creative work, arrive to the appreciation of an order in nature which is not man made or artificial but natural, contingent (we might add created) and therefore, independent of the observer.

Planck

Max (Karl Ernst Ludwig) Planck was born in Kiel from a family with a long tradition as lawyers, public servants, and university professors. He died in 1947 at Gottingen, shortly after the end of the Second World War. He then saw his dear fatherland defeated, half destroyed and in ruins. One of the most tragic moments of his life had been the trial and shooting of his son Erwin in 1944 after he had taken part in a plot against Hitler's life.

Planck studied Physics at the University of Munich. He chose Physics rather than Classical Philology or Music, disciplines for which he was equally gifted and to which he was also attracted, because he saw something very general and attractive in the physical laws governing nature. After three years in Munich, Planck moved to Berlin, where Helmholtz and Kirchhoff were professors. His doctoral dissertation centered on the Second Principle of Thermodynamics. But his attempt to discuss it with Clausius was unsuccessful.

He moved in 1885 from *Privat dozent* in Munich to Kiel, and then to Berlin, where he became Full Professor (Ordinary Professor) in 1892. At that time he became interested in the problem of the emission of radiation by a blackbody in equilibrium at a temperature T. Experimental work made then in the Physicalische-Technische-Reichsanstalt (Berlin-Charlottenburg) showed a clear connection between the intensity of the emitted radiation and the radiation wavelength, which was totally independent of the material making up the blackbody emitter. Planck saw in this, quite correctly, an absolute character of the emitted spectral distribution (specified solely by the temperature of the blackbody emitter), a clear indication that he was confronting a first class natural phenomenon, something which could well give the observer a great opportunity to go deep into the absolute character of the physical laws governing nature. He saw in it something analogous to the well known fundamental laws of thermodynamics: the law of conservation of energy and the law of increase of entropy (disorder) in closed systems.

After strenuous efforts to describe black body radiation, Planck was capable of formulating the principles of Quantum Theory which were able of explaining satisfactorily the spectral distribution of black bodies in the whole range of wave lengths. This prepared the way to explain a great variety of physical phenomena, from the photoelectric effect, to the specific heat of solids, to the absorption and emission spectra of hydrogen atoms. This pioneering work would trigger in a few decades the formulation of Quantum Mechanics.

And very important, Planck's work introduced for the first time the concept of "universal constant", beginning with Planck's own quantum of action (h) to which he immediately added the velocity of light (c), Boltzmann's constant (k_B), Newton's gravitational constant (G) and a few others.

In his investigation of blackbody radiation Planck aimed at relating this physical phenomenon to the second Law making an interpolation between two partial laws capable of describing respectively blackbody emission in the high frequency (low wave – length) limit and in the low frequency limit, consistent with the experimental results of Rubens and Kurlbaum. At the beginning Planck considered his new radiation law as "lucky intuition" and proceeded to investigate in depth its physical meaning. In his words, after a few weeks of strenuous work "the darkness lifted and an unexpected vista begun to appear". In previous work Planck regarded always the second Law of thermodynamics as an "absolute law", as absolute as the first, admitting no exception.

Subsequently he was driven to join Boltzmann in viewing that second law as an irreducible statistical law, according to which the entropy is directly related to the probability of occurrence of a state given by the number of microstates corresponding to that given macrostate.

Planck found that to justify theoretically his interpolation describing black body radiation it was necessary to assume that the energy stored in the blackbody oscillators could not be divided indefinitely but was actually made up of a number of certain, very small, "quanta" of energy, given by "$h \cdot v$", where h was Planck's constant and v was the frequency of the oscillator.

$h = 6.55 \times 10^{-27}$ erg-sec, the "quantum" of action, was, according to Planck "the most essential point of the whole calculation".

Planck was one of the first prominent physicists to defend Einstein's 1905 theory of special relativity. He endorsed also Einstein's use of "quantum theory" to explain the photoelectric effect. Years later he persuaded Einstein to come to Berlin and join the faculty there in 1914.

In his "Scientific Autobiography" (1948) Planck points out[2] that the set of universal constants identified by him, including of course his constant h, makes possible the definition of units of mass, length, time (and temperature) which are "independent of specific bodies and substances, and necessarily keep their meaning for all times and cultures, even for extra-terrestrial and extra-human cultures, and which can be properly designated as "natural units".

Einstein

Albert Einstein was born in Ulm (Germany) in March 14, 1879, and died in Princeton (USA) in April 18, 1955.

With Copernicus, Newton, and Planck, Einstein can be rightly considered one of the very few scientists who have originated scientific revolutions in history.

He was educated in Munich. At age 16 he mastered differential and integral calculus. He failed the entrance exanimation in the Federal Institute of Technology in Zürich (Switzerland) because his knowledge in non-mathematical disciplines did not match that of the mathematical ones. Following the Principal's advice, he then obtained his diploma at the Cantonal School and then was admitted to the Federal Institute of Technology at which he obtained his diploma in 1900.

Two years later he obtained a position as third class technical expert in the patent office of Bern. Six months later he married Mileva Maric, his former classmate in Zürich. They had two sons. At age, 26, Einstein completed all the requirements to receive his

Ph. D. and began writing his first original scientific papers. 1905 was his "*annus mirabilis*": in that year he published three epoch making papers. The first on Special Relativity. The second on the Photoelectric Effect. The third on Brownian motion.

In 1909, after serving as lecturer in Bern, he was invited to be associate professor at the University of Zürich. Two years later he moved to the University of Prague as full professor. One and a half years later he became full professor at the Federal Institute of Technology in Zürich. And, in 1913, Max Planck and Walter Nerst, at the time the towering figures of German physics, came to Zürich to persuade Einstein to move, as research professor, to the University of Berlin, with full membership at the Prussian Academy of Science.

Einstein accepted in 1914. He divorced Mileva Maric, who remained in Zürich, and married his cousin Elsa in 1917.

Einstein traveled extensively all over the world in the early twenties. He often campaigned for Zionism, and Phillip Lenard and Johannes Stark (German physicists, Nobel Prize winners) attacked Einstein and his theory of relativity as "Jewish" physics. Einstein resigned his position at the Prussian Academy of Science in 1933. In that year he travelled to the United States of America, and was offered a permanent position at the Institute of Advanced Studies, Princeton.

At the beginning of the Second World War, Einstein signed a letter to President Franklin D. Roosevelt, originally drafted by Leo Szilard and Eugene Wigner, proposing the fabrication of the atomic bomb. In 1945 the atomic bomb would demonstrate its devastating effects in Hiroshima and Nagasaki.

The energy released by the fission of uranium isotopes ^{238}U given by Einstein's famous equation $E = m \cdot c^2$ was enormous, but if would be dwarfed some years later by the nuclear energy released by the fusion of hydrogen nuclei.

In 1917 Einstein introduced a simple derivation of Planck's radiation law by postulating that atoms may absorb energy always spontaneously, but they may emit energy by spontaneous emission and by stimulated emission (forced by incoming radiation). This makes possible the generation of "laser" light (Light Absorption and Stimulated Emission of Radiation) with far reaching consequences to be exploited much later, in the second half of the 20th century.

The Indian physicist Satyendra N. Bose corresponded with Einstein about the proper statistics applicable to photons (light quanta) as well as to other elementary excitations occurring in condensed matter. This gave rise to the Bose-Einstein statistics, a form of non-classical quantum statistics applicable to bosons (particles with zero rest mass and integer spins). The Fermi-Dirac statistics another non-classical quantum statistics applicable to fermions (particles with non-zero rest mass, and half integer spins, like protons, neutrons and electrons).

Luis de Broglie introduced in 1924 the revolutionary idea that material particles could behave as waves. Einstein saw immediately the connection with Bose-Einstein statistics and told Schrödinger about it. In 1926 Erwin Schrödinger wrote down his famous wave equation for material particles, the first step in formulating quantum wave mechanics, which was shown later to be equivalent to Heisenberg's matrix quantum mechanics. Therefore, the birth of Quantum Mechanics owes much to Einstein's insights in spite of the fact that he was never in full accord with the interpretation given to Quantum Mechanics by Niels Bohr and his Copenhagen school.

Einstein special theory of relativity was developed independently of famous Michelson-Morley experiment (1887), but, of course, this experiment gave full support to Einstein's postulate that the velocity of light is constant, independent of the relative movement of the emitter and the detector. That is why

Einstein affirmed that his theory should have been called the theory of "invariance" rather than the theory of relativity.

The General Theory of Relativity[3], applicable to accelerated bodies, different and much more general than the Special Theory of Relativity, is the most original contribution of Einstein to theoretical physics. General Relativity predicted the deflection of starlight by massive bodies, the gravitational redshift of light emitted by very massive bodies and the long unexplained advance of the perihelion of Mercury in his trajectory around the Sun.

When in 1919 an astronomical expedition organized by Eddington confirmed the deflection of light rays from distant stars going tangent to the Sun's surface during an eclipse, Einstein became overnight and instant celebrity all over the world.

Fr. Jaki on Planck and Einstein

In chapters 11 and 12 of "The Road of Science and the Ways to God", Fr. Jaki makes a penetrating analysis of the realistic world view common to the two great pioneers of 20th century physics. Given the tremendous scientific authority accorded to Ernst Mach in the German speaking world in the late 19th century and the early 20th century, it is very instructive to begin with an evaluation of Mach's philosophical world view. Mach read Kant's "Critique of Pure Reason" when he was fifteen and he said years later that this book "made powerful and ineffaceable impression on me the like of which I never afterwards experienced in any of my philosophical readings" ("The Analysis of Sensations and the Relation of the Physical to the Psychical", New York: Dover, 1959).

In his principal publication he said: "If the individual facts... were immediately accessible to us, science would never hadraised" ("History and Root of the Principle of Conservation of Energy "translate and annotated by Phillip E.B. Jourdain, Chicago: Open Court Publishing Company, 1911). For him, only sensory perceptions mattered. He saw in the ideographic writing of the Chinese the genuinely sensationist and economic recording of ideas. If such was the case, however, one should have expected viable science to have developed in China rather than in Medieval Christian Europe.

Mach crusade against atomism, against quantum physics and against relativistic physics shows clearly the fundamental shortcomings of his philosophical world view.

On the other hand, as Fr. Jaki notes, Ernst Mach, as rector of the University of Prague required routinely the oath of Catholic orthodoxy from the newly appointed professors every year in spite of being himself a professed freethinker and agnostic. According to Philipp Franck, a close associate of Mach, his philosophy and history of science lead him to advocating Buddhism, and to the classical doctrine of eternal returns, closely resembling Nietzsche's views.

Ernest Mach conversion to Buddhism was widely aired after his death in 1916. Frank would later found the Verein Mach in 1920, which became afterwards the Vienna Circle, the cradle of logical positivism.

Max Planck, according to Fr. Jaki, on the other hand, was very much aware of the importance of the hypothesis of the quanta for 20th century physics and said "…the hypothesis of quanta will never vanish from the world… with this hypothesis a new foundation is laid for the construction of a theory which one day is destined to permeate the swift and delicate events of the molecular world with a new light" (Planck letter to Ehrenfest, July 6, 1905, quoted in "Thermodynamics and Quanta in Planck's Work", by M. Klein, p. 32). Within a few years Bohr presented his theory of the hydrogen atom.

It took from Planck a very human event (Boltzmann's suicide in 1906) to say, quoting the Gospel, "By their fruits you shall know them" at closing his first public lecture on the Weltanschauung without which, Planck believed, physics could not exist. (As Fr. Jaki notes, Boltzmann's atomistic and statistical approach to physical reality had triggered Mach's unmerciful attack) "Atoms—declared Planck—little as we know of their actual properties are as real as heavenly bodies or as earthly objects around us". In this respect, Planck referred to the estimate of the mass of Neptune long before it was seen as to a scientific procedure which made no sense within Mach's "sensationism".

After knowing his son Erwin's execution at the end of the Second World War, Planck wrote these memorable words to a friend: "What helps me is that I consider it a favor of heaven that since childhood a faith is planted deep in my innermost being, a faith in the Almighty and the All-good not to be shattered by anything. Of course his ways are not our ways, but trust in him helps us through the darkest trials" (Hermann, "Max Planck in Selbstzengnissen und Bilddokumenten p. 88).

As noted by Fr. Jaki, Albert Einstein opening reference in his contribution to Maxwell's commemorative volume might have been written by Planck himself: "the belief in an external world independent of the perceiving subject is the basis of all natural science" ("Clerk Maxwell's Influence on the Evolution of the Idea of Physical Reality", in "The World as I See It", New York: Covici – Friede 1934, p. 60).

According to Einstein "even if a physical theory appears to be doing justice to the perceived facts in the most logically perfect way", it could not be final because "sense perception gives information of the external world or physical reality only indirectly. In other words, if sense perceptions are an ongoing process, the possibility of new information remains ever present, which is an obvious source of incompleteness". Thus, according to

Einstein the status of final truth could not be accorded even to a theory like Maxwell's, which had an inclusive character and was carefully justified by his author on the basis of very general considerations.

Later, in his Spencer Memorial Lecture, Einstein gave unambiguous evidence that he viewed the whole development of science through his own creative experience, which entailed a realistic metaphysics and a realistic epistemology and was as far away from pure idealism as it was from positivism. ("On the Method of Theoretical Physics", in "The World as I See It", p.30)

According to Fr. Jaki, Einstein, who said he did not believe in a personal God, held high moral ideals with authoritative content. Their source was no reason, he said "Their highest form was deposited in the Jewish-Christian religious tradition" ("Out of My Latter Years", New York: Philosophical Library, 1950, pp. 26-27).

The world's views of Planck and Einstein, the two towering pioneers of Modern Physics, as shown by Fr. Jaki, were and are perfectly compatible with a well done and orderly natural world and with the recognition in man's intellect of a genuine capability to understand it.

True Science is a **monumental proof** of the fact that **the universe** is **well made and contingent**, and that **so is man's intellect**. Their existence requires therefore a necessary, all powerful and intelligent Creator.

REFERENCES

[1] Stanley L. Jaki, "The Road of Sciencie and the Ways to God" (The University of Chicago Press: Chicago, 1978) (Reprinted by Real View Books: Port Huron, MI, 2006).

[2] Max Planck, "Wissenschafttsliche Selbsbiographie" (FriedrVieweg & Sons: Braunschweig, 1958)

[3] Albert Einstein, "The Meaning of Relativity" (Princeton University Press: Princeton, NJ, 1922).

Ernst Mach, "The Analysis of Sensations and the Relation of the Physical to the Psychical" (Dover: NewYork, 1959)

2. ABOUT THE ORIGINS OF QUANTUM MECHANICS I

Niels Bohr
Arnold Sommerfeld

Bohr[1]

The Danish physicist Niels Henrik David Bohr (1885-1962) formulated the first successful explanation of some major lines of the hydrogen spectrum. The Bohr theory of the atom has become the foundation of modern atomic physics.

Niels Bohr (pronounced *bor*) was born on Oct. 7, 1885, in Copenhagen, the son of Christian Bohr and Ellen Adler Bohr. He studied physics and philosophy at the University of Copenhagen. His postgraduate work culminated in 1911 in a doctoral dissertation on the electron theory of metals. In the same year he

---

[1]By Stanley L. Jaki.

went to Cambridge University and worked with J. J. Thompson at the Cavendish Laboratory. By the spring of 1912 he was working with Ernest Rutherford at the University of Manchester. It was there that Bohr made some valuable suggestions about the chemical relevance of radioactive decay which proved to be most instrumental in formulating the concept of isotopes.

Secret of the Atom

Bohr's principal interest lay, however, in the planetary model of the atom, which Rutherford proposed in 1911. While pondering the implications of that model, Bohr became acquainted with Johannes Rydberg's studies of spectral lines and with J. J. Balmer's formula. As Bohr himself recalled in 1934, "As soon as I saw Balmer's formula the whole thing was immediately clear to me." The "thing" was the recognition on Bohr's part that basically different laws govern the atom when it is not in its stationary state but is absorbing or emitting radiation. He was no longer at Rutherford's laboratory when he succeeded in developing this revolutionary notion into a consistent and concise picture of the atom.

Meanwhile, in 1912 Bohr married Margrethe Norlund shortly after his return to Copenhagen, where he was appointed assistant professor at the university.

When Bohr asked Rutherford to recommend his now historic paper "On the Constitution of Atoms and Molecules" for publication, Rutherford admitted that Bohr's ideas as to the mode of origin of the spectra of hydrogen were very ingenious and worked very well, but he was unwilling to agree with Bohr's own evaluation of the paper. It took a special trip by Bohr to Rutherford in Manchester and a series of evenings during which the two carefully went over every paragraph in the paper before Rutherford's objections could be overcome. When the paper was published, in three parts in the *Philosophical Magazine*, June, September, and November 1913, reactions were divided. Some immediately expressed unreserved admiration, but there were doubters as well. In Einstein's eyes the paper was one of the great discoveries.

Copenhagen School

Bohr spent two years with Rutherford before returning to Copenhagen, where he began to think that the most effective cultivation of atomic and nuclear physics demanded a special institute, sheltering not only a well equipped laboratory, but also playing host to a large number of physicists from all over the world. In 1917 he approached the university with his plan, and as soon as the war was over the plan was enthusiastically approved. The institute was financed by public subscription, and the city donated a choice site to the Institute for Theoretical Physics, which soon established itself as the world center of theoretical physics.

Bohr's first major scientific award was the Hughes Medal of the Royal Society in 1921. The Nobel Prize followed the next year, but the finest tribute to Bohr was the steady stream of brilliant young physicists to his institute, which was dedicated on Sept. 15, 1920. Among the first to arrive at Bohr's institute was Wolfgang Pauli, and two years later, in 1924, came Werner Heisenberg, and shortly afterward Paul Dirac, to mention only some most important names in modern physics. In fact, there was hardly a major theoretical breakthrough in physics in the 1920s without some connection with the so-called Copenhagen school. Heisenberg's matrix mechanics, Erwin Schrodinger's wave mechanics, the demonstration of their equivalence by Max Born, Dirac, and P. Jordan, Pauli's theory of electron spin, Louis de Broglie's wave theory of matter-all entered the mainstream of physics through the animated discussions at Bohr's institute. Reminiscing on the 1920s, Bohr could rightly say that "in these years a unique cooperation of a whole generation of theoretical physicists from many countries created step by step, a logically consistent generalization of quantum mechanics and electromagnetics, and has sometimes been designated as the heroic age in quantum physics."

Principle of Complementarity

To use Bohr's own words, "a new outlook emerged" which put the comprehension of physical experience into radically new perspectives. Bohr contributed an important part to that new outlook when he outlined his principle of complementarity in 1927. According to Bohr, waves and particles were two complementary aspects of nature which, as far as human perception and reasoning went, represented mutually irreducible aspects of nature. The wider implications of such an outlook were further articulated by Bohr in subsequent years, as he came to grips with such philosophical questions as indeterminism versus causality, and life versus mechanism.

Bohr's famous extension of the principle of complementarity to the question of life versus mechanism came in 1932 in a lecture entitled "light and life." In this lecture he first pointed out that an exhaustive investigation of the basic units of life was impossible because those life units would most likely be destroyed by the high-speed particles needed for their observation. For Bohr, the units of life represented irreducible entities similar to the quantum of energy. According to him, the "essential nonanalyzability of atomic stability in mechanical terms presents a close analogy to the impossibility of a physical or chemical explanation of the peculiar functions characteristic of life." Scientists who, because of the subsequent startling developments in molecular biology, claimed to have come to the threshold of a mechanistic explanation of life found no ally in Bohr. To the end of his life he held fast to the basic message of his now-classic lecture, as may be seen from his essay "Light and Life Revisited", written in 1962, the year he died. An even more fundamental aspect of the principle of complementarity was the recognition that the observer and the observed represented a continuous interaction in which the two influenced and altered one another, however slightly. This meant that the rigid line of separation between the subjective and the objective needed some modification. This also meant a radical modification of the physicist's concept of the external world. The impact of the new insight into the correlation of the objective and the subjective was enormous also on the philosophical temper of the age. It seems indeed that the enunciation of the principle of complementarity by Bohr produced an insurmountable stumbling block for a mechanistic or reductionist explanation of the realm of reality as it is conceived and experienced by man.

Compound Nucleus and the Fission Process

With the discovery of the neutron in 1932, attention rapidly turned from electrons, which form the outer part of the atom, to the nucleus. To understand the various phenomena produced when nuclei of atoms were exposed to bombardment by neutrons, physicists first turned to Bohr's atom model. There the electrons moved largely independently of one another and were subject mainly to a field of force that was the average effect of the motion and position of all of them. The great number of nuclear resonances seemed, however, to point toward a rather different situation. The recognition of this came from Bohr himself, who proposed in 1936 that the protons and neutrons in the nucleus should be considered as a strongly coupled system of particles, in a close analogy to molecules making up a drop of water. In such a system there had to be a very large number of resonance levels of energy, and it also followed that a fairly long time could elapse before the available energy would concentrate on a single neutron resulting in its emission.

This picture of the "compound nucleus" formed the basis of Bohr's other crucial contribution to nuclear physics, the analysis of the fission process. In a paper written jointly with John A. Wheeler in 1939, he showed in quantitative detail the behavior of the compound nucleus for the cases of radiation, neutron emission, and fission. On this last point their all-important contribution consisted in arguing that in the fission of uranium it was mainly the isotope U^{235} that produced the effect under the impact of slow neutrons. It then became immediately clear that to obtain either a large-scale or a sustained, low-rate energy process by fissioning uranium, one had to achieve a separation of U^{235} in sufficient quantities from uranium ore in which the non fissionable U^{238} was predominant.

A Towering Figure

After 1939 Bohr's life was largely devoted to humanitarian efforts, such as intervening for the Danish Jews; he had to save human lives including his own and those of his family. Moreover, he felt duty- bound to prevent science from turning into a tool of wholesale destruction. Following his escape to Sweden in September 1943, he was quickly flown to England and from there to the United States. There he lent his talents to the Manhattan Project and during his stay at Los Alamos he did work on the initiator phase of the activation of the atomic bomb. He also began to stress the need for international control of atomic weapons and energy. His view and arguments helped shape the Acheson-Lilienthal plan and the Baruch proposals to the United Nations on behalf of the American government. In 1950 he submitted in a letter to the United Nations a plea for an "open world where each nation can assert itself solely by the extent to which it can contribute to the common culture, and is able to help others with experience and resources." In the 1950s Bohr's principal contribution to science consisted in taking a leading part in the development of the European Center for Nuclear Research (CERN). It was at his institute that the decision was made to build the 28-Bev (billion-electron-volt) accelerator near Geneva.

From 1938 until his death he was the president of the Royal Danish Academy of Sciences, acted as chairman of the Danish Atomic Energy Commission, and supervised the first phase of the Commission's program for the peaceful uses of atomic energy.

Bohr's last major appearance was to deliver the Rutherford Memorial Lecture in 1961 which gave a fascinating portrayal not only of the great master but also of his equally famous disciple.

Bohr's death came rather suddenly but quietly on Nov. 18, 1962, at his home. Einstein and he were possibly the most towering and influential figures of 20th-century physics.

Further Reading

Niels Bohr is listed in the *Science Study Guide* (VII, C, 2). Among the notables who attended Bohr's institute were Wolfgang Pauli, Werner Heisenberg and Paul Dirac.

The best biography of Bohr is Ruth Moore, *Niels Bohr: The Man, His Science and the World They Changed* (1966). Stefan Rozental, ed. *Niels Bohr: His Life and Work as Seen by His Friends and Colleagues* (trans. 1967), is a most valuable collection of essays contributed by Bohr's closest friends and associates. On Bohr's role in 20th-century physics one should consult the papers written in his honor on his seventieth birthday, W. Pauli, ed., *Niels Bohr and the Development of Physics* (1955). See also Niels Hugh de Vaudrey Heathcote, *Nobel Prize Winners in Physics*, 1901-1950 (1953); Arthur March and Ira Freeman, *New World of Physics* (1962), and Henry A. Boorse and Lloyd Motz, ed., *The World of the Atom* (2 vols., 1966).

Arnold Sommerfeld[2]

Arnold Johannes Wilhelm Sommerfeld (5 December 1868-26 April 1951) was a leading German physicist and a pioneer of the development of atomic physics and quantum physics. He had more Nobel prize winners among his students than any other Ph.D. thesis supervisor at any time in the last century.

He introduced a second quantum number (the azimuthal quantum number) in Bohr's model of hydrogen-like atoms to improve the description of energy levels, and then a fourth quantum number (the spin).

---

[2]By Julio A. Gonzalo

The famous "finite structure constant" $\alpha = \dfrac{e^2}{\hbar c} \cong \dfrac{1}{137}$ (dimensionless) was introduced by him to characterize the strength of the electromagnetic interaction between charged particles. (At the time, Eddington, the great British theoretician became so much obsessed with α that, entering a public establishment, he looked for the number 137 to hang his hat on it).

Also he was a pioneer of the theory of X-ray diffraction.

He studied at Konigsberg (Prussia), taught at Gottingen, Clausthat, Aachen and Munich, where he retired there as Professor Emeritus in 1951. His thesis advisor was Professor von Lindemann.

The list of his doctoral students is truly impressive: W. Heisenberg, W. Pauli, P. Debye, P.S. Epstein, H. Bethe, E. Guillemin, K. Bechart, P.P. Ewald, H. Frölich, E. Fues, H. Hönl, L. Hopf, W. Kossel, A. Kratzer, A, Landé, O. Laporte, W. Lenz, R. Pierls, W. Rogowski, R. Seeliger, H. Welker, G. Wentzel.

At a time when German Science (physics, chemistry, engineering…) was at its most creative period, Sommerfeld's contributions were many and important. In particular:

1. Sommerfeld combined the classical Drude theory of electrons in metals with the quantum mechanical statistical theory of Fermi-Dirac to develop successfully the modern theory of electrons in metals. The free electron empty lattice approximation forms the basis for the band structure model known as the nearly free electron model. This theory succeeds in the explanation of phenomena such as the Wiedemann – Franz Law (connecting electron conductivity with thermal conductivity in metals). Sommerfeld's work, in collaboration with his gifted graduate students, made important contributions to explain the temperature dependence of the

specific heat of solids, the behavior of the electronic density of states, the order of magnitude of the binding energy of compounds, the electrical conductivity of solids and the thermoelectric emission of bulk metals.

2. As mentioned before, Sommerfeld introduced the "fine structure constant" (1916). Its numerical value $\alpha = 7.29735257 \times 10^{-3} \cong 1/137$ presently determined with very high precision, both theoretically and experimentally, is one of the most spectacular successes of Quantum Electro Dynamics. Experimentally its precise quantitative determination is possible thanks to the quantum Hall effect. The magnetic moment of the electron (Lande factor) and the fine structure constant (α) are directly related.

If one defines $q_{Planck} \equiv (\hbar/c)^{1/2}$, α can be interpreted as the squared ratio between the electron charge (e) and Planck's charge (q_{Pl}).

3. Sommerfeld identity. Is a mathematical identity about the propagation of spherical waves. Its physical interpretation is that a spherical wave can be expanded into a summation of cylindrical waves in the axial direction of the cylinder multiplied by a plane wave in the perpendicular direction.

4. The so called Old Quantum Theory due to Bohr and Sommerfeld is a collection of semi-empirical results (1900-25) which predates the rigorous formulation of Quantum Mechanics by Schrödinger and Heisenberg. Sommerfeld made an important contribution when he introduced into Bohr's primitive theory for the hydrogen atom the quantization of the z component of the angular momentum. This allowed the orbits of the electrons to become ellipses instead of circumferences by introducing the idea of quantum degeneracy. The Old Quantum Theory of Sommerfeld and

Bohr could have explained the Zeemann effect except for complications due to the electron spin.

For a more detailed review of Sommerfeld's contributions, see f.i. "Arnold Johannes Wilhelm Sommerfeld 1868-1951". *Obituary Notices of Fellows of the Royal Society.*

3. ABOUT THE ORIGINS OF QUANTUM MECHANICS II

Louis de Broglie
Erwin Schrodinger
Werner Heisenberg
Pascual Jordan

De Broglie[1]

Prince Louis-Victor de Broglie of the French Academy, Permanent Secretary of the Academy of Sciences, and Professor at the Faculty of Sciences at Paris University, was born at Dieppe (Seine Inferieure) on 15th August, 1892, the son of Victor, Duc de Broglie and Pauline d'Armaille. After studying at the Lycee Janson of Sailly, he passed his school-leaving certificate in 1909. He applied himself first to literary studies and took his degree in

---

[1] From *Nobel Lecture, Physics 1922-1941*, Elsevier Publishing Company, Amsterdam, 1965. This autobiography/biography was written at the time of the award and later published in the book series *Les Prix Nobel/Nobel Lectures*.

MLA style: "Louis de Broglie - Biography". Nobelprize.org. 7 Dec 2010. http://nobelprize.org/nobel_prizes/physics/laureates/1929/broglie.html

history in 1910. Then, as his liking for science prevailed, he studied for a science degree, which he gained in 1913. He was then conscripted for military service and posted to the wireless section of the army, where he remained for the whole of the war of 1914-1918. During this period he was stationed at the Eiffel Tower, where he devoted his spare time to the study of technical problems. At the end of the war Louis de Broglie resumed his studies of general physics. While taking an interest in the experimental work carried out by his elder brother, Maurice, and co-workers, he specialized in theoretical physics and, in particular, in the study of problems involving quanta. In 1924 at the Faculty of Sciences at Paris University he delivered a thesis *Recherches sur la Theorie des Quanta* (Researches on the Quantum Theory), which gained him his doctor's degree. This thesis contained a series of important findings which he had obtained in the course of about two years. The ideas set out in that work, which first gave rise to astonishment owing to their novelty, were subsequently fully confirmed by the discovery of electron diffraction by crystals in 1927 by Davisson and Germer; they served as the basis for developing the general theory nowadays known by the name of *wave mechanics*, a theory which has utterly transformed our knowledge of physical phenomena on the atomic scale.

After the maintaining of his thesis and while continuing to publish original work on the new mechanics, Louis de Broglie took up teaching duties. On completion of two year's free lectures at the Sorbonne he was appointed to teach theoretical physics at the Institut Henri Poincare which had just been built in Paris. The purpose of that Institute is to teach and develop mathematical and theoretical physics. The incumbent of the chair of theoretical physics at the Faculty of Sciences at the University of Paris since 1932, Louis de Broglie runs a course on a different subject each year at the Institut Henri Poincare, and several of these courses have been published. Many French and foreign students have come to work with him and a great deal of doctorate theses have been prepared under his guidance.

Between 1930 and 1950, Louis de Broglie's work has been chiefly devoted to the study of the various extensions of wave mechanics:

Dirac's electron theory, the new theory of light, the general theory of spin particles, applications of wave mechanics to nuclear physics, etc. He has published numerous notes and several papers on this subject, and is the author of more than twenty-five books on the fields of his particular interests.

Since 1951, together with young colleagues, Louis de Broglie has resumed the study of an attempt which he made in 1927 under the name of the *theory of the double solution* to give a causal interpretation to wave mechanics in the classical terms of space and time, an attempt which he had then abandoned in the face of the almost universal adherence of physicists to the purely probabilistic interpretation of Born, Bohr, and Heisenberg. Back again in this his former field of research, he has obtained a certain number of new and encouraging results which he has published in notes to *Comptes Rendus de 1 'Academie des Sciences* and in various expositions.

After crowning Louis de Broglie's work on two occasions, the Academie des Sciences awarded him in 1929 the Henri Poincare medal (awarded for the first time), then in 1932, the Albert I of Monaco prize. In 1929 the Swedish Academy of Sciences conferred on him the Nobel Prize for Physics "for his discovery of the wave nature of electrons". In 1952 the first Kalinga Prize was awarded to him by UNESCO for his efforts to explain aspects of modern physics to the layman. In 1956 he received the gold medal of the French NationalScientific Research Centre. He has made major contributions to the fostering of international scientific cooperation.

Elected a member of the Academy of Sciences of the French Institute in 1933, Louis de Broglie has been its Permanent

Secretary for the mathematical sciences since 1942. He has been a member of the Bureau des Longitudes since 1944. He holds the Grand Cross of the Legion d'Honneur and is an Officer of the Order of Leopold of Belgium. He is an honorary doctor of the Universities of Warsaw, Bucharest, Athens, Lausanne, Quebec, and Brussels, and a member of eighteen foreign academies in Europe, India, and the U.S.A.

Louis de Broglie died on March 19, 1987.

Professor De Broglie's most important publications are:

Recherches sur la theorie des quanta (Researches on the quantum theory), Thesis Paris, 1924.
Ondes et mouvements (Waves and motions), Gauthier-Villars, Paris, 1926.
Rapport au 5e Conseil de Physique Solvay, Brussels, 1927.
La mecanique ondulatoire (Wave mechanics), Gauthier-Villars, Paris, 1928.
Une tentative d'interpretation causale et non lineaire de la mecanique ondulatoire: la theorie de la double solution, Gauthier-Villars, Paris, 1956.
English translation: *Non-linear Wave Mechanics: A Causal Interpretation*, Elsevier, Amsterdam, 1960.
Introduction a la nouvelle theorie des particules de M. Jean-Pierre Vigier et de ses collaborateurs, Gauthier-Villars, Paris, 1961.
English translation: *Introduction to the Vigier Theory of elementary particles*, Elsevier, Amsterdam, 1963.
Etude critique des bases de l'interpretation actuelle de la mecanique ondulatoire, Gauthier-Villars, Paris, 1963.
English translation: *The Current Interpretation of Wave Mechanics: A Critical Study*, Elsevier, Amsterdam, 1964.

Schrodinger[2]

Erwin Schrodinger was born on August 12, 1887, in Vienna, the only child of Rudolf Schrodinger, who was married to a daughter of Alexander Bauer, his Professor of Chemistry at the Technical College of Vienna.

Erwin's father came from a Bavarian family which generations before had settled in Vienna. He was a highly gifted man with a

[2]From *Nobel Lectures.Physics 1922-1941*, Elsevier Publishing Company, Amsterdam, 1965. This autobiography/ biography was written at the time of the award and later published in the book series *Les Prix Nobel/Nobel Lectures*. It was later edited and republished in *Nobel Lectures*. MLA style: "Erwin Schrodinger - Biography". Nobelprize.org. 7 Dec 2010.

broad education. After having finished his chemistry studies, he devoted himself for years to Italian painting. After this he took up botany, which resulted in a series of papers on plant phylogeny.

Schrodinger's wide interests dated from his school years at the Gymnasium, where he not only had a liking for the scientific disciplines, but also appreciated the severe logic of ancient grammar and the beauty of German poetry. (What he abhorred was memorizing of data and learning from books.)

From 1906 to 1910 he was a student at the University of Vienna, during which time he came under the strong influence of Fritz Hasenohrl, who was Boltzmann's successor. It was in these years that Schrodinger acquired a mastery of eigenvalue problems in the physics of continuous media, thus laying the foundation for his future great work. Hereafter, as assistant to Franz Exner, he, together with his friend K. W. F. Kohlrausch, conducted practical work for students (without himself, as he said, learning what experimenting was).

During the First World War he served as an artillery officer. In 1920 he took up an academic position as assistant to Max Wien, followed by positions at Stuttgart (extraordinary professor), Breslau (ordinary professor), and at the University of Zurich (replacing von Laue) where he settled for six years. In later years Schrodinger looked back to his Zurich period with great pleasure - it was here that he enjoyed so much the contact and friendship of many of his colleagues, among whom were Hermann Weyl and Peter Debye. It was also his most fruitful period, being actively engaged in a variety of subjects of theoretical physics. His papers at that time dealt with specific heats of solids, with problems of thermodynamics (he was greatly interested in Boltzmann's probability theory) and of atomic spectra; in addition, he indulged in physiological studies of colour (as a result of his contacts with Kohlrausch and Exner, and of Helmholtz's lectures).

His great discovery, Schrodinger's wave equation, was made at the end of this epoch-during the first half of 1926. It came as a result of his dissatisfaction with the quantum condition in Bohr's orbit theory and his belief that atomic spectra should really be determined by some kind of eigenvalue problem. For this work he shared with Dirac the Nobel Prize for 1933.

In 1927 Schrodinger moved to Berlin as Planck's successor. Germany's capital was then a centre of great scientific activity and he enthusiastically took part in the weekly colloquies among colleagues, many of whom "exceeding him in age and reputation". With Hitler's coming to power (1933), however, Schrodinger decided he could not continue in Germany. He came to England and for a while held a fellowship at Oxford. In 1934 he was invited to lecture at Princeton University and was offered a permanent position there, but did not accept. In 1936 he was offered a position at University of Graz, which he accepted only after much deliberation and because his longing for his native country outweighed his caution. With the annexation of Austria in 1938, he was immediately in difficulty because his leaving Germany in 1933 was taken to be an unfriendly act. Soon afterwards he managed to escape to Italy, from where he proceeded to Oxford and then to University of Ghent. After a short stay he moved to the newly created Institute for Advanced Studies in Dublin, where he became Director of the School for Theoretical Physics. He remained in Dublin until his retirement in 1955.

All this time Schrodinger continued his research and published many papers on a variety of topics, including the problem of unifying gravitation and electromagnetism, which also absorbed Einstein and which is still unsolved; (he was also the author of the well-known little book *What is Life?*, 1944). He remained greatly interested in the foundations of atomic physics. Schrodinger disliked the generally accepted dual description in terms of waves and particles, with a statistical interpretation for the waves, and tried to set up a theory in terms of waves only. This led him into controversy with other leading physicists.

After his retirement he returned to an honoured position in Vienna. He died on the 4th of January, 1961, after a long illness, survived by his faithful companion, Annemarie Bertel, whom he married in 1920.

Heisenberg[3]

Werner Heisenberg (pronounced hi'zan-berkh) was born on Dec. 5, 1901, in Wurzburg, the son of August and Annie Wecklein Heisenberg. He received his education at the Maximilian Gymnasium in Munich and at the University of Munich, where his father was professor of Greek language and literature. Shortly before he began his university studies, he worked on a farm for several months and took active part in youth movements searching

---

[3]By Stanley L. Jaki.

for a way out of the social collapse that hit Germany at the end of World War I. He was also a talented pianist, an avid hiker, and an eager student of classical literature and philosophy. At the university, where he matriculated, in 1920, Heisenberg soon established close contact with Arnold Sommerfeld, a chief figure in early modern physics, and with Sommerfeld's most outstanding student, Wolfgang Pauli, a later Nobel laureate. Heisenberg spent the winter of 1922/1923 at the University of Gottingen, where the physics department was rapidly establishing itself, with the help of Max Born, James Franck, and David Hilbert, as a center of theoretical physics. After taking his doctorate in Munich in 1923, Heisenberg went on a Rockefeller grant to Niels Bohr's institute in Copenhagen, where he eagerly studied the most creative and up-to-date speculations on atomic theory.

His Landmark Papers

The fusion of best influences with the receptiveness of a most talented mind worked unusually well. No sooner had Heisenberg completed his stay in Copenhagen than he worked out, while recuperating on the shores of Helgoland from a heavy attack of hay fever, a comprehensive method of calculating the energy levels of "atomic oscillators". The method yielded very good results but appeared so strange that Heisenberg was undecided whether to submit his report for publication or "to throw it into the flames". Happily for science, he sent a copy of it to Pauli, and after receiving a favorable reply from him, he showed it on his return to Gottingen in June 1925 to Born. Born immediately recognized its importance and had it sent to the *Physikalische Zeitschrift*, where it was immediately printed under the title "On Quantum Mechanical Interpretation of Kinematic and Mechanical Relations." The most-preoccupied with the *strange* mathematical formalism in Heisenberg's paper was Born himself, who after eight days of constant reflection discovered that it corresponded to the rules of matrix calculus.

Heisenberg's paper earned its author immediate fame and recognition. At Bohr's recommendation, in 1926 he was appointed lecturer in theoretical physics at the University of Copenhagen. It was there that Heisenberg gave much thought to the apparent discrepancy between two formulations of quantum theory, one based on matrix calculus, the other on wave equations elaborated by Erwin Schrodinger. In the course of his work on this question, Heisenberg realized that only those physical situations are *meaningful* in quantum mechanics in which the differences of the non commutative products of conjugate variables occur. He immediately saw that because of these differences one cannot determine simultaneously the position and velocity of an atomic particle or the energy level and its timing of an atomic oscillator. The recognition of this led Heisenberg to the formulation of the famous uncertainty principle, which appeared in 1927 on the pages of the *Physikalische Zeitschrift* in an article entitled "On the Visualizable Content of Quantum Theoretical Kinematics and Mechanics".

Academic Career and Books

Heisenberg's rise was now as rapid in the academic as in the scientific world. In 1927, at the age of 26, he became professor of theoretical physics at the University of Leipzig. He was the recipient with Schrodinger and Paul Dirac of the Nobel Prize for physics for 1932. In 1941 he took the chair of theoretical physics at the University of Berlin and the directorship of the Kaiser Wilhelm Institute for Physics. During the war he worked as an enlisted officer on the German uranium project and was in Allied captivity in England from April 1945 until the summer of 1946. After returning to Germany he did much to reorganize scientific research as head of the Max Planck Institute of Physics and of the Alexander von Humboldt Foundation.

In the early 1950s Heisenberg turned with great vigor toward the formulation of a "unified theory of fundamental particles" stressing the role of symmetry principles. This theory was intensively discussed at an international conference in 1958, the year he moved to the University of Munich as professor of physics. He presented his thought on this subject in *Introduction to the Unified Field Theory of Elementary Particles* (1966).

Heisenberg's *The Physical Principles of the Quantum Theory* (1930) is a classic in the field. In 1955-1956 he gave the Gifford Lectures at the University of St. Andrews, Scotland, which were printed under the title *Physics and Philosophy: The Revolution in Modern Science*. He also published the autobiographical *Physics and Beyond* (1971) and several books dealing with the philosophical and cultural implications of atomic and nuclear physics, all of which are available in English translation. In 1937 he married Elisabeth Schumacher, and they had seven children.

Further Reading

Werner Heisenberg is listed in the *Science Study Guide* (VII, C, 1). He collaborated with Max Born.

The best treatment of the conceptual foundations of Heisenberg's achievements in physics is the study by Patrick A. Heelan, *Quantum Mechanics and Objectivity: A Study of the Physical Philosophy of Werner Heisenberg* (1965). The place of Heisenberg's discoveries in the development of modern physics is given with all the technical details in the work by Max Jammer, *The Conceptual Development of Quantum Mechanics* (1966): For a popular but still informative presentation of the origins and techniques of quantum mechanics see Banesh Hoffmann, *The Strange Story of the Quantum* (1959). For an account sprinkled with anecdotal details see the works of George Gamow, *Biography of Physics* (1961) and *Thirty Years That Shook Physics: The Story of Quantum Theory* (1966).

Jordan[4]

Pascual Jordan (1902-1980), one of the truly great physicists of the 20th century, was founder, together with Max Born and Werner Heisenberg, of Quantum Mechanics, and together with Wolfgang Pauli and Eugene Wigner, of Quantum Electrodynamics. He is the only one of the five great physicists mentioned above who was not awarded the Nobel Prize in Physics. His pre-war sympathies with the Nazi party in Germany may have had something to do with it. He was the originator of the quantum theory of fields and did theoretical work in cosmology, assuming that Newton's

[4]By Julio A. Gonzalo

gravitational constant G was time dependent and was decreasing like the reciprocal of the age of the universe.

His uncommon first name is owed to a Spanish great-grandfather who had served with Napoleon's army and remained in Germany after the great retreat from Moscow.

He came to the conclusion that all things in the universe - photons, electrons, protons, atoms and even macroscopic entities- are field quanta. He, with Max Born and Werner Heisenberg, formulated quantum mechanics in the famous *Dreimännerarbeit* of 1925, but he never received the same acclaim as his colleagues.

He was considered rather a "formalist", meaning that he was not a true physicist, but only a mathematician. His formalism was apt to contain the physical truth in quantum physics. Max Born said in the 1950s: "In December 1925 I went to America to give lectures at MIT. I was editor of *Zeitschrift für Physik* and Jordan gave me a paper for publication in the journal. I didn't find time to read it and put it in my suitcase and forgot all about it. Then, when I came back home to Germany half a year later and unpacked, I found the paper at the button of the suitcase. It contained what came to be known as the Fermi-Dirac Statistics. In the mean time it had been discovered by Enrico Fermi and independently, by Paul Dirac. But Jordan was the first".

In spite of his pre-war political proclivities, Jordan was personally a shy and kind man according to people who worked with him. In 1953, thanks to Pauli's support, Jordan was rehabilitated. He advanced from visiting to full professor at Hamburg.

Jordan did also work in cosmology, trying to generalise Friedmann solution to Einstein equation for the cosmic radius R(t) by treating the universe as a hyper sphere filled with matter and a time dependent scalar field, equivalent to a variable gravitational "constant". Some years before Dirac had entertained the same idea from numerical considerations connecting the immense ratio of electrical to gravitational attraction in the hydrogen atom with the ratio of the age of the universe ($\sim 4\times10^{17}$ sec) to a characteristic nuclear time ($\sim 10^{-23}$ sec). Jordan's was an attempt to turn into field theory Dirac's numerology.

In 1979, Wigner proposed Jordan for the Nobel Prize in Physics. The award was given that year to Sheldon Glashow, Abdus Salam and Steven Weinberg, three distinguished practitioners of the kind of quantum field theory invented by Jordan.

Further reading

For further reading on Pascual Jordan see "Jordan, Pauli, Politics, Brecht and a Variable Gravitational Constant" by E. L. Schucking, in *Physics Today* 52, n°. 10 (Oct-1999) and references therein.

4. ABOUT THE ORIGINS OF QUANTUM MECHANICS III

Paul Dirac
Max Born
Peter Debye
Arthur H. Compton

Dirac[1]

Paul Dirac was born on Aug. 8, 1902, at Monk Royal in Bristol, the son of Charles Adrien Ladislas Dirac and Florence Hannah Holten Dirac. Paul received his secondary education at the old Merchant Venturers' College and, at the age of 16, entered Bristol University. He graduated three years later in electrical engineering.

---

[1] By Stanley L. Jaki

Unable to find employment, he studied mathematics for two years before moving to Cambridge as research student and recipient of an 1851 Exhibition scholarship award. His student years (1923-1926) at Cambridge saw the emergence of the mathematical formulation of modern atomic physics in the hands of Louis de Broglie, Werner Heisenberg, Erwin Schrodinger, and Max Born. It was therefore natural that Dirac's attention should turn to a cultivation of mathematics most directly concerned with atomic physics.

Negative Kinetic Energy

Dirac's first remarkable contribution along these lines came before he earned his doctorate in 1926. In his paper "The Fundamental Equations of Quantum Mechanics" (1925), Dirac decided to extricate the fundamental point in Heisenberg's now famous paper. Before Heisenberg, computation of energy levels of optical and X-ray spectra consisted in a somewhat empirical extension of rules provided by Niels Bohr's theory of the atom. Heisenberg succeeded in grouping terms connected with energy levels in columns forming large squares and also indicated the marvelously simple ways in which any desired energy level could be readily calculated. Dirac found that what Heisenberg really wanted to achieve consisted in a most general type of operation on a "quantum variable" x which was done by "taking the difference of its Heisenberg products with some other quantum variable."

At that time neither Heisenberg nor Dirac had realized that the "Heisenberg products" corresponded to operations in matrix calculus, a fact which was meanwhile being proved by Born and Pascual Jordan in Gottingen. They showed that the noncommutative multiplication of the "Heisenberg quantities" could be summed up in the formula

$$(p \times q) \cdot (q \times p) = h / 2\pi\sqrt{-1}$$

Where h is Planck's constant and p and q some canonically conjugate variables. Independently of them, Dirac also obtained the same formula, but through a more fundamental approach to the problem. Dirac's crucial insight consisted in finding that a very simple operation formed the basis of the formula in question. What had to be done was to calculate the value of the classical Poisson bracket $[p,q]$ for p and q and multiply it by a modified form of Planck's constant.

That such a procedure yielded the proper values to be assigned to the difference of $(p \times q)$ and $(q \times p)$ was only one aspect of the success. The procedure also provided an outstanding justification of the principle of correspondence, tying into one logical whole the classical and modern aspects of physics. Dirac once remarked that the moment of that insight represented perhaps the most enthralling experience in his life. But the most startling result of Dirac's equation for the electron was the recognition of the possibility of negative kinetic energy. In other words, his equations implied for the electron an entirely novel type of motion whereby energy had to be put into the electron in order to bring it to rest. The novelty was both conceptual and experimental and received a remarkably quick elucidation.

The experimental clarification came when C.D. Anderson, doing cosmic-ray research in R. A. Millikan's laboratory in Pasadena, Calif., obtained on Aug. 2, 1932, the photograph of an electron path, the curvature of which could be accounted for only if the electron had a positive charge. The positively charged electron, or positron, was, however, still unconnected with the negative energy states implied in Dirac's theory of the electron. The work needed in this respect was largely done by Dirac, though not without some promptings from others. A most lucid summary of the results was given by Dirac in the lecture which he delivered on

Dec. 12, 1933, in Stockholm, when he received the Nobel Prize in physics jointly with Schrodinger.

World of Antimatter

The most startling consequences of Dirac's theory of the electron consisted in the opening up of the world of antimatter. Clearly, if negative electrons had their counterparts in positrons, it was natural to assume that protons had their counterparts as well. Here Dirac argued on the basis of the perfect symmetry that according to him had to prevail in nature. As a matter of fact, it was a lack of symmetry in Schrodinger's equation for the electron that Dirac tried to remedy by giving it a form satisfactory from the viewpoint of relativity.

All this should forcefully indicate that Dirac was a thinker of most powerful penetration who reached the most tangible conclusions from carrying to their logical extremes some utterly abstract principles and postulates. Thus by postulating the identity of all electrons, he was able to show that they had to obey one specific statistics. This fact in turn provided the long-sought clue for the particular features of the conduction of electricity in metals, a problem with which late classical physics and early quantum theory grappled in vain. This attainment of Dirac paralleled a similar, though less fundamental, work by Enrico Fermi, so that the statistics is now known as the Fermi-Dirac statistics. This contribution of Dirac came during a marvelously creative period in his life, from 1925 to 1930. Its crowning conclusion was the publication of his *Principles of Quantum Mechanics*, a work still unsurpassed for its logical compactness and boldness. The latter quality is clearly motivated by Dirac's unlimited faith in the mathematical structuring of nature. The book is indeed a monument to his confidence that future developments will provide the exact physical counterparts that some of his mathematical symbols still lack.

A telling measure of Dirac's main achievements in physics was the recognition that greeted his work immediately. In 1932 he was elected a fellow of the Royal Society and given the most prestigious post in British science, the Lucasian chair of mathematics at Cambridge. He received the Royal Society's Royal Medal in 1939 and its Copley Medal in 1952. He was a member of many academies, held numerous honorary degrees, and was a guest lecturer in universities all over the world. He married Margaret Wigner, sister of Nobel laureate Eugene P. Wigner, in 1937.

Further Reading

Paul Dirac is listed in the *Science Study Guide* (VII, C, 1). Modern atomic physics was formulated by Niels Bohr, Louis de Broglie, Erwin Schrodinger, Werner Heisenberg, Max Born, and Dirac. Enrico Fermi and Dirac developed quantum statistics.

Humorous details on Dirac's life can be found in George Gamow, *Biography of Physics* (1961), together with a not too technical discussion of Dirac's theory of holes. See also Niels H. de V. Heathcote, *Nobel Prize Winners in Physics*, 1901-1950 (1954). For a rigorous account of Dirac's role in quantum mechanics, the standard work is Max Jammer, *The Conceptual Development of Quantum Mechanics* (1966). Background works which discuss Dirac include James Jeans, Physics and Philosophy (1942), and Barbara Lovett Cline, *The Questioners: Physicists and the Quantum Theory* (1965).

Born[2]

On Dec. 11, 1882, Max Born was born in Breslau. He studied at the universities of Breslau, Heidelberg, and Zurich before he settled in Gottingen.

In accordance with the advice of his father, Born did not specialize but attended courses in the humanities as well as in the sciences, especially mathematics. In Gottingen he followed with great enthusiasm the lectures in astronomy by Karl Schwarzschild

---

[2]By Stanley L. Jaki.

but found no stimulation in the physics courses. He earned his doctorate with a dissertation in applied mathematics, namely, the analysis of the stability of elastic wires and tapes.

Although Born was inducted for the one-year compulsory military service, because of his asthmatic condition he obtained an early discharge. He went to Cambridge but within a few months returned to Breslau. In the fall of 1908 Born was back at Gottingen, where he later obtained the post of Privatdozent (lecturer) in physics on the merits of his paper on the relativistic aspects of the electron. This was the start of his career as a physicist.

Born's first outstanding achievement in physics came in 1912, when in collaboration with T. von Karman he worked out the theoretical explanation of the whole range of the variation of specific heat in solids. Although the official credit for this major feat went to Peter Debye, who independently did the same work a few weeks earlier, the topic became very decisive in Born's future work as a physicist. It opened to him the two main lines of his subsequent research: lattice dynamics and quantum theory.

In 1912 Born made his first trip to the United States to lecture on relativity at the University of Chicago. On his return to Gottingen he married Hedwig Ehrenberg; they had two children. Born's close relationship with Albert Einstein began in 1915, when Born went to the University of Berlin to take over some of the teaching duties of Max Planck. There Born's 5-year-long investigation of the dynamics of crystal lattices was published as his first book. Between 1919 and 1921 he was at the University of Frankfurt am Main.

In 1921 Born succeeded Debye at Gottingen as director of the physics department. The work of Wolfgang Pauli, Werner Heisenberg, and Erwin Schrodinger produced the major advances in quantum theory, but it was Born who reduced these various efforts to a basic foundation. It consisted in showing that the square of the value of Schrodinger's psi function was the probability density in configuration space. This meant that quantum mechanics allowed only a statistical interpretation of events on the atomic level. The result was so fundamental and startling that such leaders of modern physics as Planck, Einstein, Louis de Broglie, and Schrodinger could not bring themselves to accept it unreservedly. Born attributed to their reluctance the fact that he did not receive the Nobel Prize until three decades later, in 1954.

Born further elaborated the implications of his major discovery in his guest lectures at the Massachusetts Institute of Technology in the winter of 1925/1926, the text of which appeared under the title *Problems of Atomic Dynamics*, probably the first monograph on quantum mechanics. Born's return to Gottingen signaled the beginning of a pilgrimage of young American physicists to Gottingen. His own work, however, became handicapped by nervous exhaustion in 1928. He therefore gave up research on atomic theory and wrote a textbook on optics, which became a classic in the field. In May 1933 he had to depart from Germany only a few months after Hitler came to power. Following a short stay in South Tirol, the Borns went to Cambridge, where he concentrated on writing two books that also became classics: *The Atomic Theory* (1935) and *The Restless Universe* (1936), the latter a popular exposition. In 1936 the Borns went to India at the invitation of Sir C. V. Raman, a Nobel laureate physicist, but half a year later they were at Edinburgh, where Born succeeded Charles G. Darwin as professor of natural philosophy (physics).

Born stayed in Edinburgh seventeen years, and his major achievements there are embodied in three books: one on the lattice dynamics of crystals, a new enlarged version of his textbook on optics, and *Natural Philosophy of Cause and Chance*. The last represented the text of his Waynflete Lectures at Magdalen College, Oxford. It shows Max Born at his philosophical best, for he retained all his life a keen interest in the deeper aspects of physics. This also explains his well-known concern about the ethical implications of science and about the role of science in the general fabric of human culture.

Born had already taken up residence in Bad Pyrmont near Gottingen in 1954 when he began to publish his startling articles on these topics. His view of the future was rather dim, though he pleaded for the attitude of "hoping against hope." He died in Gottingen on Jan. 5, 1970.

Further Reading

Max Born is listed in the *Science Study Guide* (VII, C, 1). His theory of the quantum mechanics of the motion of an atomic particle, which was formulated with the assistance of Werner Heisenberg, was developed independently in England by Paul Dirac.

An invaluable source on Born is the autobiographical *My Life and My Views* (1968), which contains priceless episodes from his life, an authoritative discussion of the genesis of his principal contributions to physics, and his reflections on the role of science in modern culture. A popular but informative discussion of the development of quantum mechanics is Banesh Hoffmann, *The Strange Story of the Quantum* (1947; 2d ed. 1959). Max Jammer, *The Conceptual Development of Quantum Mechanics* (1966), will probably be for many years the standard work on the topic.

Debye[3]

Peter Debye was born on March 24, 1884, in Maastricht, Netherlands, the son of William and Maria Reumkens Debye. At the age of 17 Debye entered the Technical Institute of Aachen and earned his diploma in electrical engineering in 1905. He immediately obtained the position of assistant in technical mechanics at the institute. At the same time his interest in physics received strong promptings from Arnold Sommerfeld, then serving

---

[3]By Stanley L. Jaki

on the faculty. Debye followed Sommerfeld to the University of Munich and obtained his doctorate in physics by a mathematical analysis of the pressure of radiation on spheres of arbitrary electrical properties.

The dissertation and a 1907 paper on Foucault currents in rectangular conductors gave clear evidence of Debye's ability to produce the mathematical tools demanded by his topics. A fitting recognition of Debye's youthful excellence was his succession in 1911, at the age of 27, to Albert Einstein in the chair of theoretical physics at the University of Zurich. While in Zurich he worked out, on the basis of Max Planck'sand Einstein's ideas, the first complete theory of the specific heat of solids and the equally important theory of polar molecules. Debye was professor of theoretical physics at the University of Utrecht from 1912 until 1914, when he received the prestigious post of director of the theoretical branch of the Institute of Physics at the University of Gottingen. In 1915 he became editor of the famed *Physikalische Zeitschrift* and served in that capacity for twenty-five years.

X-ray Research

In Gottingen, Debye started a most fruitful collaboration with P. Scherrer. Their first paper, "X-ray Interference Patterns of Particles Oriented at Random" (1916), gave immediate evidence of the enormous potentialities of their powder method to explore the structure of crystals with very high symmetry. It also proved very useful in work with polycrystalline metals and colloidal systems. Two years later Debye and Scherrer extended the method from the study of the coordination of atoms to the arrangement of electrons inside the atom. It was in this connection that they formulated the important concept of *atomic form factor*. Debye and Scherrer formed such a close team that when, in 1920, Debye became professor of experimental physics and director of the physics laboratory at the Swiss Federal Technical Institute in Zurich, Scherrer followed him there. The two inaugurated a most influential X-ray research center which attracted students from all over the world.

In the field of X-ray research Debye's signal success in Zurich was his demonstration in early 1923 that in the collision between X-rays and electrons, energy and momentum are conserved; he also suggested that the interaction between electromagnetic radiation and electrons must therefore be considered as a collision between photons and electrons. But Debye's principal achievement in Zurich consisted in the formulation of his theories of magnetic cooling and of interionic attraction in electrolyte solutions. The latter work, in which he collaborated with E. Huckel, was closely related to Debye's pioneering research on dipole moments. Debye had already been for two years the director of the Physical Institute at the University of Leipzig when his classic mono-graph, *Polar Molecules*, was published in 1928.

War and Postwar Years

Debye's rather rapid moves from one university to another were motivated by his eagerness to work with the best available experimental apparatus. Thus in 1934 he readily accepted the invitation of the University of Berlin to serve both as professor at the university and as director of the Kaiser Wilhelm Institute. The latter establishment, now known as Max Planck Institute, was just completing, with the help of the Rockefeller Foundation a new laboratory which was to represent the best of its kind on the Continent. During his stay in Berlin, Debye became the recipient of the Nobel Prize in chemistry for 1936. It was awarded to him "for his contributions to our knowledge of molecular structure through his investigations on dipole moments and on the diffraction of X-rays and electrons in gases."

Meanwhile, the Nazi government began to renege on its original promise that Debye would not be asked to renounce his Dutch citizenship while serving as director of the Kaiser Wilhelm Institute, a post with a lifetime tenure. Shortly after World War II broke out, he was informed that he could no longer enter the laboratory of the institute unless he assumed German citizenship. As Debye refused, he was told to stay home and keep busy writing books.

But he succeeded in making his way to Italy and from there to Cornell University, which invited him to give the Baker Lectures in 1940. Debye made Cornell his permanent home. He served there as head of the chemistry department for the next ten years. His wartime service to his adopted country (he became a citizen in 1946) concerned the synthetic rubber program. In pure research he further investigated, in collaboration with his son Peter P. Debye, the light-scattering properties of polymers, on which he based the now generally accepted absolute determination of their molecular weights. He was a member of all leading scientific societies and the recipient of all major awards in chemistry. His outgoing personality kept generating enthusiasm and goodwill throughout his long life, which came to an end on Nov. 2, 1966. Since 1913 he had been married to Mathilde Alberer, who shared his lively interest in gardening and fishing.

Further Reading

Peter Debye is listed in the *Science Study Guide* (VII, D, 2). The demonstration of the conservation of energy and momentum was achieved independently in 1923 by Arthur Compton in the United States.

The best sources available on Debye's life and on the various aspects of his scientific work are the introductory essays in *The Collected Papers of Peter J. W. Debye* (1954). A detailed biographical profile of Debye is in the Royal Society, *Biographical Memoirs of Fellows of the Royal Society*, vol. 16 (1970). Debye is discussed in Eduard Farber, *Nobel Prize Winners in Chemistry*,

1901-1961 (1953; rev. ed. 1963); Aaron I. Ihde, *The Development of Modern Chemistry* (1964); and *Chemistry: Nobel Lectures, Including Presentation Speeches and Laureates' Biographies, 1922-41*, published by the Nobel Foundation (1966).

Compton[4]

Arthur Holly Compton was born at Wooster, Ohio, on September 10th, 1892, the son of Elias Compton, Professor of Philosophy and Dean of the College of Wooster. He was educated at the College, graduating Bachelor of Science in 1913, and he

---

[4]From *Nobel Lectures.Physics 1922-1941*, Elsevier Publishing Company, Amsterdam, 1965. This autobiography/ biography was written at the time of the award and later published in the book series *Les Prix Nobel*. It was later edited and republished in *Nobel Lectures*. MLA style: "Arthur H. Compton - Biography". Nobelprize.org. 7 Dec 2010.

http://nobelprize.org/nobel_prizes/physics/laureates/1927/compton.htm

spent three years in postgraduate study at Princeton University receiving his M.A. degree in 1914 and his Ph.D. in 1916. After spending a year as instructor of physics at the University of Minnesota, he took a position as a research engineer with the Westinghouse Lamp Company at Pittsburgh until 1919 when he studied at Cambridge University as a National Research Council Fellow. In 1920, he was appointed Wayman Crow Professor of Physics, and Head of the Department of Physics at the Washington University, St. Louis; and in 1923 he moved to the University of Chicago as Professor of Physics. Compton returned to St. Louis as Chancellor in 1945 and from 1954 until his retirement in 1961 he was Distinguished Service Professor of Natural Philosophy at the Washington University.

In his early days at Princeton, Compton devised an elegant method for demonstrating the Earth's rotation, but he was soon to begin his studies in the field of X-rays. He developed a theory of the intensity of X-ray reflection from crystals as a means of studying the arrangement of electrons and atoms, and in 1918 he started a study of X-ray scattering. This led, in 1922, to his discovery of the increase of wavelength of X-rays due to scattering of the incident radiation by free electrons, which implies that the scattered quanta have less energy than the quanta of the original beam. This effect, nowadays known as the *Compton effect*, which clearly illustrates the particle concept of electromagnetic radiation, was afterwards substantiated by C. T. R. Wilson who, in his cloud chamber, could show the presence of the tracks of the recoil electrons. Another proof of the reality of this phenomenon was supplied by the coincidence method (developed by Compton and A.W. Simon, and independently in Germany by W. Bothe and H. Geiger), by which it could be established that individual scattered X-ray photons and recoil electrons appear at the same instant, contradicting the views then being developed by some investigators in an attempt to reconcile quantum views with the continuous waves of electromagnetic theory. For this discovery, Compton was awarded the Nobel Prize in Physics for 1927 (sharing this with C. T. R. Wilson who received the Prize for his discovery of the cloud chamber method).

In addition, Compton discovered (with C. F. Hagenow) the phenomenon of total reflection of X-rays and their complete polarization, which led to a more accurate determination of the number of electrons in an atom. He was also the first (with R. L. Doan) who obtained X-ray spectra from ruled gratings, which offers a direct method of measuring the wavelength of X-rays. By comparing these spectra with those obtained when using a crystal, the absolute value of the grating space of the crystal can be determined. The Avogadro number found by combining above value with the measured crystal density, led to a new value for the electronic charge. This outcome necessitated the revision of the Millikan oil-drop value from 4.774 to 4.803 x 10^{-10} e.s.u. (revealing that systematic errors had been made in the measurement of the viscosity of air, a quantity entering into the oil-drop method). During 1930-1940, Compton led a world-wide study of the geographic variations of the intensity of cosmic rays, thereby fully confirming the observations made in 1927 by J. Clay from Amsterdam of the influence of latitude on cosmic ray intensity. He could, however, show that the intensity was correlated with geomagnetic rather than geographic latitude. This gave rise to extensive studies of the interaction of the Earth's magnetic field with the incoming isotropic stream of primary charged particles.

Compton has numerous papers on scientific record and he is the author of *Secondary Radiations Produced by X-rays* (1922), *X-Rays and Electrons* (1926, second edition 1928), *X-Rays in Theory and Experiment* (with S. K. Allison, 1935, this being the revised edition of *X-rays and Electrons*), *The Freedom of Man* (1935, third edition 1939), *On Going to College* (with others, 1940), and *Human Meaning of Science* (1940).

Dr. Compton was awarded numerous honorary degrees and other distinctions including the Rumford Gold Medal (American Academy of Arts and Sciences), 1927; Gold Medal of Radiological Society of North America, 1928; Hughes Medal (Royal Society) and Franklin Medal (Franklin Institute), 1940.

He served as President of the American Physical Society (1934), of the American Association of Scientific Workers (1939-1940), and of the American Association for the Advancement of Science (1942).

In 1941 Compton was appointed Chairman of the National Academy of Sciences Committee to Evaluate Use of Atomic Energy in War. His investigations, carried out in cooperation with E. Fermi, L. Szilard, E. P. Wigner and others, led to the establishment of the first controlled uranium fission reactors, and, ultimately, to the large plutonium-producing reactors in Hanford, Washington, which produced the plutonium for the Nagasaki bomb, in August 1945. (He also played a role in the Government's decision to use the bomb; a personal account of these matters may be found in his book, *Atomic Quest - a Personal Narrative*, 1956).

In 1916, he married Betty Charity McCloskey. The eldest of their two sons, Arthur Allen, is in the American Foreign Service and the youngest, John Joseph, is Professor of Philosophy at the Vanderbilt University (Nashville, Tennessee). His brother Wilson is a former President of the Washington State University, and his brother Karl Taylor was formerly President of the Massachusetts Institute of Technology.

Compton's chief recreations were tennis, astronomy, photography and music.

He died on March 15th, 1962, in Berkeley, California.

5. ABOUT THE ORIGINS OF QUANTUM MECHANICS IV

Wolfgang Pauli
Satyendra Bose
Enrico Fermi
Eugene P. Wigner

Pauli[1]

Wolfgang Pauli was born on April 25th, 1900 in Vienna. He received his early education in Vienna before studying at the

---

[1]From Nobel Lectures. Physics 1942-1962, Elsevier Publishing Company, Amsterdam, 1964. This autobiography/biography was written at the time of the award and later published in the book series Les Prix Nobel. It was later edited and republished in Nobel Lectures. MLA style: "Wolfgang Pauli - Biography". Nobelprize.org. 7 Dec 2010.

http://nobelprize.org/nobel_prizes/physics/laureates/1945/pauli.html

University of Munich under Arnold Sommerfeld. He obtained his doctor's degree in 1921 and spent a year at the University of Gottingen as assistant to Max Born and a further year with Niels Bohr at Copenhagen. The years 1923-1928 were spent as a lecturer at the University of Hamburg before his appointment as Professor of Theoretical Physics at the Federal Institute of Technology in Zurich. During 1935-1936, he was visiting Professor at the Institute for Advanced Study, Princeton, New Jersey and he had similar appointments at the University of Michigan (1931 and 1941) and Purdue University (1942). He was elected to the Chair of Theoretical Physics at Princeton in 1940 but he returned to Zurich at the end of World War II.

Pauli was outstanding among the brilliant mid-twentieth century school of physicists. He was recognized as one of the leaders when, barely out of his teens and still a student, he published a masterly exposition of the theory of relativity. His exclusion principle, which is often quoted bearing his name, crystallized the existing knowledge of atomic structure at the time it was postulated and it led to the recognition of the two-valued variable required to characterize the state of an electron. Pauli was the first to recognize the existence of the neutrino, an uncharged and massless particle which carries off energy in radioactive β-disintegration; this came at the beginning of a great decade, prior to World War II, for his centre of research in theoretical physics at Zurich.

Pauli helped to lay the foundations of the quantum theory of fields and he participated actively in the great advances made in this domain around 1945. Earlier, he had further consolidated field theory by giving proof of the relationship between spin and "statistics" of elementary particles. He has written many articles on problems of theoretical physics, mostly quantum mechanics, in scientific journals of many countries; his *Theory of Relativity* appears in the *Enzyklopaedie der Mathematischen Wissenschaften*, Volume 5, Part 2 (1920), his *Quantum Theory* in *Handbuch der*

Physik, Vol. 23 (1926), *and his Principles of Wave Mechanics* in *Handbuch der Physik,* Vol. 24 (1933).

Pauli was a Foreign Member of the Royal Society of London and a member of the Swiss Physical Society, the American Physical Society and the American Association for the Advancement of Science. He was awarded the Lorentz Medal in 1930.

Wolfgang Pauli married Franciska Bertram on April 4th, 1934. He died in Zurich on December 15th, 1958.

Bose[2]

Satyendra Nath Bose (1894-1974) is an Indian physicist famous because he originated, together with Albert Einstein, one of the two varieties of Quantum Statistics, Bose-Einstein Statistics, applicable to one kind of quantum particles, called *bosons*, whose wave functions are *symmetric*, like electromagnetic radiation (made up of photons), lattice vibrations in solids (made up of phonons), superconductive pairs (made up of paired electrons in superconductors) etc. The other variety of Quantum Statistics, the Fermi-Dirac Statistics, is applicable to other kind of particles called "fermions", whose wave functions are *antisymmetric*.

---

[2]By Julio A. Gonzalo.

For systems at low densities and high temperatures, quantum effects are usually negligible, and Classical Statistics, i.e. Maxwell- Boltzmann Statistics applies.

In 1924 Bose, then at the Dacca University in India, wrote a letter to Einstein saying:

> Respected Sir, I have ventured to send you the accompanying article for your perusal and opinion... You will see that I have tried to deduce the coefficient $\pi v^2 / c^3$ in Planck's law independent of classical thermodynamics, only assuming that the ultimate elementary region in the phase space has the content h^3. I do not know sufficient German to translate the paper. If you think the paper worth publication I shall be grateful if you arrange for its publication in *Zeitschrift fur Physik*...

The article was published under the title "Plank's Gesetz und Lichtquantenhypothese", *Zeitschrift fur Physik*, 26, 178-181 (1924). Subsequently, Einstein using Bose's approach to the monoatomic gas, realised that a far reaching relationship between radiation and gas can be established. He showed that if the gas particles are subjected to the new statistics, the mean-square energy fluctuation is given as composed by two additive terms, one which corresponds to Maxwell- Boltzmann Statistics for non-interacting particles, and the other to interference fluctuation, associated with undulatory phenomena. Thus was born what now is known as the Bose-Einstein Statistics. Satyendra Nath Bose, whose name became even more famous after recent experiments confirming the phenomenon of Bose-Einstein condensation, died at the age of 80 in 1974. He deserves a full biography still to be written.

Fermi[3]

Enrico Fermi was born in Rome on 29th September, 1901, the son of Alberto Fermi, a Chief Inspector of the Ministry of Communications, and Ida de Gattis. He attended a local grammar school, and his early aptitude for mathematics and physics was

---

[3]From *Nobel Lectures. Physics* 1922-1941, Elsevier Publishing Company, Amsterdam, 1965. This autobiography/biography was written at the time of the award and later published in the book series *Les Prix Nobel.* It was later edited and republished in *Nobel Lectures.* MLA style: "Enrico Fermi - Biography". Nobelprize.org. 7 Dec 2010.

http://nobelprize.org/nobel_prizes/physics/laureates/1938/fermi.html

recognized and encouraged by his father's colleagues, among them A. Amidei. In 1918, he won a fellowship of the Scuola Normale Superiore of Pisa. He spent four years at the University of Pisa, gaining his doctor's degree in physics in 1922, with Professor Puccianti.

Soon afterwards, in 1923, he was awarded a scholarship from the Italian Government and spent some months with Professor Max Born in Gottingen. With a Rockefeller Fellowship, in 1924, he moved to Leyden to work with P. Ehrenfest, and later that same year he returned to Italy to occupy for two years (1924-1926) the post of Lecturer in Mathematical Physics and Mechanics at the University of Florence.

In 1926, Fermi discovered the statistical laws, nowadays known as the "Fermi statistics", governing the particles subject to Pauli's exclusion principle (now referred to as *fermions*, in contrast with *bosons* which obey the Bose-Einstein statistics).

In 1927, Fermi was elected Professor of Theoretical Physics at the University of Rome (a post which he retained until 1938, when he - immediately after the receipt of the Nobel Prize- emigrated to America, primarily to escape Mussolini's fascist dictatorship).

During the early years of his career in Rome he occupied himself with electrodynamic problems and with theoretical investigations on various spectroscopic phenomena. But a capital turning-point came when he directed his attention from the outer electrons towards the atomic nucleus itself. In 1934, he evolved the β-decay theory, coalescing previous work on radiation theory with Pauli's idea of the neutrino. Following the discovery by Curie and Joliot of artificial radioactivity(1934), he demonstrated that nuclear transformation occurs in almost every element subjected to neutron bombardment. This work resulted in the discovery of slow neutrons that same year, leading to the discovery of nuclear fission and the production of elements lying beyond what was until then the Periodic Table.

In 1938, Fermi was without doubt the greatest expert on neutrons, and he continued his work on this topic on his arrival in the United States, where he was soon appointed Professor of Physics at Columbia University, N.Y. (1939-1942).

Upon the discovery of fission, by Hahn and Strassmann early in 1939, he immediately saw the possibility of emission of secondary neutrons and of a chain reaction. He proceeded to work with tremendous enthusiasm, and directed a classical series of experiments which ultimately led to the atomic pile and the first controlled nuclear chain reaction. This took place in Chicago on December 2, 1942 -on a squash court situated beneath Chicago's stadium. He subsequently played an important part in solving the problems connected with the development of the first atomic bomb. (He was one of the leaders of the team of physicists on the Manhattan Project for the development of nuclear energy and the atomic bomb).

In 1944, Fermi became American citizen, and at the end of the war (1946) he accepted a professorship at the Institute for Nuclear Studies of the University of Chicago, a position which he held until his untimely death in 1954. There he turned his attention to high-energy physics, and led investigations into the pion-nucleon interaction.

During the last years of his life Fermi occupied himself with the problem of the mysterious origin of cosmic rays, thereby developing a theory, according to which a universal magnetic field -acting as a giant accelerator- would account for the fantastic energies present in the cosmic ray particles.

Professor Fermi was the author of numerous papers both in theoretical and experimental physics. His most important contributions were: "Sulla quantizzazione del gas perfetto monoatomico", *Rend. Accad. Naz. Lincei*, 1935 (also in Z. Phys., 1936), concerning the foundations of the statistics of the electronic gas and of the gases made of particles that obey the Pauli Principle.

Several papers published in *Rend. Accad. Naz. Lincei*, 1927-28, deal with the statistical model of the atom (Thomas-Fermi atom model) and give a semi quantitative method for the calculation of atomic properties. A resume of this work was published by Fermi in the volume: *Quantentheorie und Chemie*, edited by H. Falkenhagen, Leipzig, 1928. "Uber die magnetischen Momente der AtomKerne", *Z. Phys.*, 1930, is a quantitative theory of the hyperfine structures of spectrum lines. The magnetic moments of some nuclei are deduced therefrom.

"Tentativo di una teoria dei raggi β ", *Ricerca Scientifica*, 1933 (also *Z. Phys.*, 1934) proposes a theory of the emission of β -rays, based on the hypothesis, first proposed by Pauli, of the existence of the neutrino. The Nobel Prize for Physics was awarded to Fermi for his work on the artificial radioactivity produced by neutrons, and for nuclear reactions brought about by slow neutrons. The first paper on this subject "Radioattivita indotta dal bombardamento di neutroni" was published by him in *Ricerca Scientifica*, 1934. All the work is collected in the following papers by himself and various collaborators: "Artificial radioactivity produced by neutron bombardment", *Proc. Roy. Soc.*, 1934 and 1935; "On the absorption and diffusion of slow neutrons", *Phys. Rev.*, 1936. The theoretical problems connected with the neutron are discussed by Fermi in the paper "Sul moto dei neutroni lenti", *Ricerca Scientfica*, 1936.

His *Collected Papers* are being published by a Committee under the Chairmanship of his friend and former pupil, Professor E. Segre (Nobel Prize winner 1959, with O. Chamberlain, for the discovery of the antiproton).

Fermi was member of several academies and learned societies in Italy and abroad (he was early in his career, in 1929, chosen among the first 30 members of the Royal Academy of Italy).

As lecturer he was always in great demand (he has also given several courses at the University of Michigan, Ann Arbor; and Stanford University, Calif.). He was the first recipient of a special award of $50,000 -which now bears his name- for work on the atom.

Professor Fermi married Laura Capon in 1928. They had one son Giulio and one daughter Nella. His favourite pastimes were walking, mountaineering, and winter sports.

He died in Chicago on 29th November, 1954.

Wigner[4]

Eugene Paul Wigner, born in Budapest, Hungary, on November 17, 1902, naturalized a citizen of the United States on January 8, 1937, has been since 1938 Thomas D. Jones Professor of

---

[4] From *Nobel Lectures, Physics* 1963-1970, Elsevier Publishing Company, Amsterdam, 1972. This autobiography/biography was written at the time of the award and first published in the book series *Les Prix Nobel*. It was later edited and republished in *Nobel Lectures*. MLA style: "Eugene Wigner - Biography". Nobelprize.org.7 Dec 2010.

http://nobelprize.org/nobel_prizes/physics/laureates/1963/wigner.html

Mathematical Physics at Princeton University -he retired in 1971. His formal education was acquired in Europe; he obtained the Dr. Ing. degree at the Technische Hochschule Berlin. Married in 1941 to Mary Annette Wheeler, he is the father of two children, David and Martha. His son, David, is teaching mathematics at the University of California in Berkeley. His daughter, Martha, is with the Chicago area transportation system, an organization endeavoring to improve the internal transportation system of that city.

Dr. Wigner worked on the Manhattan Project at the University of Chicago during World War II, from 1942 to 1945, and in 1946-1947 became Director of Research and Development at Clinton Laboratories. Official recognition of his work in nuclear research includes the U. S. Medal for Merit, presented in 1946; the Enrico Fermi Prize (U.S.A.E.C.) awarded in 1958; and the Atoms for PeaceAward, in 1960. Dr. Wigner holds the Medal of the Franklin Society, the Max Planck Medal of the German Physical Society, the George Washington Award of the American-Hungarian Studies Foundation (1964), the Semmelweiss Medal of the American-Hungarian Medical Association (1965), and the National Medal of Science (1969). He has received honorary degrees from the University of Wisconsin, Washington University, Case Institute, University of Alberta (Canada), University of Chicago, Colby College, University of Pennsylvania, Yeshiva University, Thiel College, Notre Dame University, Technische Universitat Berlin, Swarthmore College, Universite de Louvain, Universite de Liege, University of Illinois, Seton Hall, Catholic University and The Rockefeller University. He is a past vice-president and president of the American Physical Society, of which he remains a member. He is a past member of the board of directors of the American Nuclear Society and still a member; he holds memberships in the American Philosophical Society, the American Mathematical Society, the American Association of Physics Teachers, the National Academy of Science, the American Academy of Arts and Sciences, the Royal Netherlands Academy of Sciences and Letters, the

American Association for the Advancement of Science, the
Austrian Academy of Sciences, he is corresponding member of the
Gesellschaft der Wissenschaften, Gottingen, and foreign member
of the Royal Society of Great Britain. He was a member of the
General Advisory Committee to the U.S. Atomic Energy
Commission from 1952-1957, was reappointed to this committee
in 1959 and served on it until 1964. Eugene Wigner died on
January 1, 1995.

6. INDETERMINACY VS UNCERTAINTY

Indeterminacy in nature means, according to the Copenhagen interpretation of Quantum Mechanics that nature "itself" is somehow "undecided", and therefore "free" to decide or not to decide. Let us quote Bohr, Heisenberg's principal mentor, in this connection[1]: "one speaks of a free choice on the part of nature". He added that this phrase could be seen as implying "the idea of an external chooser", but he discarded it right away, to avoid giving the impression that he could consider seriously any genuinely transcendental Creator.

On the other hand, **Uncertainty** means something very different. It simply means the lack of certainty in one's knowledge about natural reality or about aspects of that natural reality. At the final roundtable of the International Summer Course on "Astrophysical Cosmology: Frontier Questions"[2] held at El Escorial, near Madrid, in 1993, it came up the question of **indeterminacy** vs. **uncertainty** in Quantum Mechanics, and Ralph A. Alpher, who had given the opening lecture at the Summer Course, voted clearly in favor of uncertainty. In fact, Heisenberg's principle says simply that the product of the uncertainties of two complementary conjugate variables is restricted to be more than $\hbar$ (Planck's constant divided by 2π). In other words

$$\Delta E \cdot \Delta t \geq \hbar \tag{6.1}$$

(E = energy, t = time),

$$\Delta p \cdot \Delta x \geq \hbar \qquad (6.2)$$

(p = momentum, x = spatial coordinate),

Is it purely negative the information content of Heisenberg's principle?

The right answer is that not at all: First, $\hbar$ is extremely small ($\hbar = 1.05 \times 10^{-27}$ ergs $\cdot$ sec) and, therefore, for any macroscopic, mesoscopic or even ordinary microscopic system (like a proton, a neutron, an electron or a positron) the uncertainty in the pair of conjugated variables (energy and time; momentum and spatial coordinate) is small or relatively moderate; Second, taking into consideration that the uncertainty in mass, spatial or temporal extension of any given finite physical system can be reasonably assumed to be at most as large as the physical quantity itself (known by other means) one can get very valuable information about that physical quantity by means of Heisenberg's uncertainty principle.

Consider the almost negligible mass of the neutrino emitted in the β-decay of a radioactive nucleus.

The dimensionless quantity[3] giving the relative strength of the nuclear weak interaction is

$$\alpha_{NW} \cong \frac{g_W^2 e^{-r/r''}}{\hbar c} \cong 10^{-12} \qquad (6.3)$$

($g_W^2 / \hbar c \cong 15 \times 10^{-12}$, $r'' \cong 1.5 \times 10^{-15}$ cm)

Therefore $\Delta E \cdot \Delta r \cong \alpha_{NW} \hbar c$, $\qquad (6.4)$

($\hbar = 1.05 \times 10^{-27}$ erg $\cdot$ sec, $c = 3 \times 10^{10}$ cm/sec

Resulting for a neutrino of negligible mass

$$(m_v c^2)(r''/c) \cong 10^{-12} \hbar, \tag{6.5}$$

Which implies

$$m_v \cong 2.2 \times 10^{-35} \text{ g} \cong 2.95 \bullet 10^{-5} m_e, \tag{6.6}$$

negligible in comparison with the electron mass, and certainly insufficient to make the cosmic density parameter

$$\Omega_m = \rho_m / \rho_{mc} = 1 \quad \text{(the cosmic density parameter)}$$

Therefore, Heisenberg's uncertainty principle gives very useful order of magnitude information about such an elusive microscopic physical particle as the electron neutrino mass.

It may be recalled that when Fermi noted that, apparently, some energy was missing in any β decay event, Bohr was quick to say that may be in the microscopic realm of atoms, nuclei and elementary particles, energy conservation should not be taken for granted. It was Pauli who suggested in 1930 the possible existence of the neutrino, a neutral particle of negligible mass, to insure energy conservation.

In 1956, Clyde Cowan and Frederick Reines (who was awarded the Physics Nobel Prize many years after the death of Cowan) demonstrated beautifully the existence of neutrinos bombarding pure water with a beam of 10^{18} neutrinos per second in a nuclear reactor.

They used the reaction[4]

$$\bar{v} + {}'H' \rightarrow {}^o n' + \bar{e} \tag{6.7}$$

where $\bar{v}$ stands for the neutrino, ${}^o n'$ for the neutron and $\bar{e}$ for the positron, the inverse of which is the alternative of the neutron decay

$$^{\circ}n^{'} \rightarrow {}^{'}H^{'} + \bar{e} + \bar{\nu} \tag{6.8}$$

The Reines-Cowan reaction took place in the hydrogen atoms of a very large scintillation counter subject to the enormous flux of neutrinos from the fission induced β decays in a nuclear reactor.

The uncertainty resulting from the dual wave-particle character of zero rest mass photons (as discovered by Compton), and material particles with non-zero rest mass (as suggested by de Broglie and demonstrated for electrons very soon by the Davisson-Germer experiment), can be taken as a restriction on the observers ability to measure exactly. An observer who is always macroscopic. As Eddington noted[5], *molar physics has always the last word in observation, for the observer himself is molar*. There is no doubt that Quantum Mechanics, as formulated by Werner Heisenberg, Max Born, and Pascual Jordan, or, equivalently, by Erwin Schrodinger, works very well to describe events in the microscopic world. Heisenberg's uncertainty principle is fully justified. But how about the general assumption of indeterminacy (a strictly philosophical assumption), an assumption that implies that we are prohibited to consider any underlying objective reality which, on the other hand we must necessarily take for granted, for obvious reasons, even if our knowledge of that reality is limited?

The uncertainty of our knowledge is a fact. The indeterminacy of "nature" is an arbitrary artificial assumption.

For Bohr (a follower of Mach's extreme positivistic philosophy of nature) only "sensations" are real. Concepts, judgments and logically connected reasoning are not grounded in any objective reality). For Einstein (a realist: "the belief in an external world independent of the perceiving subject is the basis of all natural science"). Bohr apparently won, but Einstein was not alone. With him were none others than Planck, de Broglie, Compton, and Schrodinger. Let us quote P. Ehrenfest and P. Dirac on Bohr's[6] extreme positivistic philosophy.

P. Ehrenfest: *Teaching Bohr's mathematical physics "in its present confusion[1]" gave him (Ehrenfest) insight into Hegelian dialectics – "a succession of leaps from one lie to another by way of intermediate falsehoods."*

P. Dirac: *The experiments of W. Bothe and H. Geiger (1924) showed[1] that the coincidence of scattered X-rays and recoil electrons made wholly untenable a theory proposed by Bohr, Kramers and Slater which held that atomic processes were **purely statistical** and, therefore, that the principles of energy and momentum conservation were not applicable. Regarding Bohr's denial of the validity of "trajectories" on elementary particles, Dirac pointed out that no facts were then more obvious than the ionization tracks shown in cloud chambers. Nothing was more natural for him than allowing for a **causal mechanism** of condensation from one nucleus to another. On the basis of a purely probabilistic quantum-mechanical basis, the invariable appearance of well-defined tracks in cloud chambers had to appear a great puzzle* (as well documented in Ref (1)).

In any event, it is fair to say that calling Heisenberg's principle the **principle of uncertainty** reflects better its true **meaning**. That denomination, however, does not reflect quite well its positive potential to provide very useful information about a finite physical system.

REFERENCES

[1]Stanley L. Jaki, "The Road of Science and the Ways to God" (The University

of Chicago Press: Chicago, 1978) p. 203.

[2] Julio A. Gonzalo, Jose L. Sanchez Gomez and Miguel A. Alario (eds), "Cosmología Astrofísica" (Alianza Universidad: Madrid 1996).

[3] Julio A. Gonzalo, "Cosmic Paradoxes" (World Scientific: Singapore, 2012).

[4] Robert Eisberg and Robert Resnick, "Quantum Physics of Atoms, Molecule, Solids, Nuclei and Particles" (John Wiley & Sons: New York, London, Sydney, Toronto, 1974).

[5] A. Eddington, "The Philosophy of Physical Science" (Macmillan Company: New York, 1939).

[6] Stanley L. Jaki, Ibidem. Pp. 197-213

7. THE UNIVERSAL CONSTANTS

Heisenberg's uncertainty principle has an intrinsic connection with the universal constant $\hbar$, Planck's quantum of action divided by 2π

$$\hbar = 1.05 \times 10^{-27} \text{ erg} \cdot \text{sec} \tag{7.1}$$

On April 11th, 2000, at 12:30 pm an academic session was held at the main Conference Room of the Faculty of Science, Universidad Autónoma de Madrid (UAM) to celebrate the 1st Centennial of **Planck's Quantum of Action**. That session was opened by Professor Rodolfo Miranda, Vice Chancellor for Research, on behalf of Professor Emilio Crespo, Vice Chancellor for Cultural Activities. He pointed out that Planck's discovery would change the way scientists perceive physical reality in the future by his preparing the way for the uncertainty principle. The next Speaker, Professor Agustín Gárate, Vice Dean of Professorate of the Faculty of Sciences, said that as teachers and students of Physical Sciences (especially Physics and Chemistry) we know very well the tremendous impact of Planck's work in the twentieth century. Planck's work and Planck's constant (h) lead very soon to the understanding of the photoelectric effect (Einstein), the specific heat of solids (Einstein and Debye) and the atomic spectra of the hydrogen atom (Bohr). Then I introduced the Invited Lecturer for this occasion, Professor Stanley L. Jaki, Distinguished Professor at Seton Hall University, South Orange, New Jersey, Templeton Prize, 1987. Professor Jaki had been at UAM in 1992, when he gave a very interesting Lecture on "Is there such a thing as a last word in Physics?", and then at 1997, when he spoke about "Lucky coincidences at the Earth Moon System and their relevance for Drake's equation". At the Academic Session he entitled his lecture "Numbers decideor Planck's constant and some constants of philosophy", quoting Plank's words at his Nobel Lecture, June 2, 1920, at which he tried to give in his own words the story of the origin of quantum theory in broad outline, its development up to that time, and its significance for physics.

In his first illustration, Professor Jaki showed a picture with Planck's tombstone at the cemetery of Gottingen in which Planck's effigy and his name appear together with his quantum's numerical expression,

$$h = 6.67 \times 10^{-27} \text{ erg} \cdot \text{sec}$$

Then he proceeded to do a detailed and masterful analysis of the ups and downs of Planck's discovery of the "quantum of action" as a result of his laborious interpretation of the "black body" emission in the whole range of wave lengths. He showed convincingly the validity of Planck's summary of his epoch making discovery in just two words "**Numbers decide**". He showed[1] first Wienn's formula which described very well black body radiation at low wave lengths

$$\phi_\lambda = \frac{C}{\lambda^5} e^{-c/\lambda\theta} \text{ (Wienn, 1896)} \tag{7.2}$$

Where C and c are constants, λ is the radiation wavelength and θ the absolute temperature. And then he showed Planck's formulae, in quick succession, leading to the description of black body radiation at any wave length by

$$E = \frac{8\pi ch}{\lambda^5} \frac{1}{e^{ch/k\lambda\theta} - 1} \text{ (Planck, Jan 7, 1901)} \tag{7.3}$$

The expression "black body" or "black cavity" radiation was yet to come into wide use. Since those black body cavities give a minimum light at a given temperature, they were used as instruments of calibration for light sources (including light bulbs) which at that time were coming into wide use all over the place.

Wienn's derivation of his displacement law (1897) which registered the displacement towards lower wave lengths of the emission maxima, called widespread attention, and his work was quickly translated into English and published in the "Philosophical Magazine", the most prestigious British journal for physics at the time. Wienn had relied for his derivation on the Stephan – Boltzmannlaw.

In November 1899 Lummer and Pringsheim, and then Rubens, had published new results showing that when the temperature of the blackbody emitter was increased there appeared deviations from Wienn's Law (7.2) at high wave lengths. In October 7, 1900, Rubens visited Planck and in the evening of that day Planck was quick to write Rubens telling him that he had just obtained a new law which agreed very well with the new data. A few days later, on October 19, Planck presented his new formula to the Physikalische Gesellschaft including all recent data and stressing the point that the **numbers** were **decisive** in confirming what he saw as a **fundamental law of nature**.

Max Born, recalling a conversation[2] with Lummer and Pringsheim, wrote in 1906: "Although Planck's formula was in the center of discussion, one was nevertheless inclined to view Planck's proposal of **quantum like oscillator energy** only as a provisional working hypothesis, and Einstein's light quanta were not taken seriously".

On February 27, 1909, Planck wrote to Wienn (a good friend from his youth years) dismissing the radical inference of Einstein about the propagation of light in the form of **bundles** or **quanta** of energy.

On another occasion Planck's recalled Wienn's hunting skills as well as his skills as a physicist[3]: "There are but few physicists who mastered so equally as Willy Wienn did the experimental and the theoretical parts of their science, and in the future it will happen on even fewer occasions that one and the same researcher would make so different discoveries as the displacement law of black-body radiation and the nature of Cathode rays"

Even in 1910 Planck wrote: "The introduction of the quantum of action *h* into the theory should be done as **conservatively** as possible, i.e., alterations should only be made that have shown themselves to be absolutely necessary". When in 1911 Wienn received the Nobel Prize for Physics, no reference was made to the fact that Wienn's formula had been shown to be incorrect in the infrared range. In his acceptance speech Wienn both praised and criticized Planck, and suggested that one should look into Sommerfeld's interpretation of h to get a physical meaning. At that time Bohr's atom model was still not published.

In a paper presented by Planck at the Solvay Conference, Brussels, October 1911, where all the leading "progressive" physicists were present, he insisted that his hypothesis was not an "Energie-hypothese" but a "Wirkungshypothese"[4]. In any case Planck brought to a conclusion his speech on that occasion as follows[5]:

> "The largest part of the work is still to be done. Surely some death flowers will keep falling from the tree of knowledge. But the beginning has already been made. The hypothesis of the quantum shall not disappear from the world. The laws of heat radiation guarantee this. And I do not think to go too far when I state my opinion that through that hypothesis the foundation is laid for the construction of a theory which is destined to cast in a new light the fast and delicate moving events of the molecular world".

The concept of universal constant must be traced back to Planck's theoretical work on blackbody radiation[6]. His quantum theory was successful to account for

(a) The spectral distribution of black body radiation form the lowest to the higher frequencies

$$W_T(\omega)\, d\omega = \frac{\hbar}{\pi} \frac{\omega^3}{c^3} \frac{1}{e^{\hbar\omega/k_B T} - 1}\, d\omega, \qquad (7.4)$$

equivalent to Eq. (7.3), where W_T is the thermal radiation energy density per unit volume and per unit frequency interval, T

the absolute temperature, ω the angular frequency, $\hbar = h/2\pi$, and k_B = Boltzmann's constant.

(b) The Wienn's displacement law, according to which the maximum of the emitted radiation for a given equilibrium temperature T occurs at a certain $\omega_{\max}$ such that it grows with T as

$$\hbar \omega_{\max} \cong 2.8 k_B T, \tag{7.5}$$

and

(c) The Stephan-Boltzmann law, which gives the total amount of emitted radiation per unit volume in the whole frequency range from zero to infinity, as

$$\int_0^\infty W_T(\omega)\, d\omega = \left[\frac{\pi^2}{15} (\hbar c)^{-3} k_B^4 \right] T^4 \equiv \sigma T^4 \tag{7.6}$$

These three relations allowed Planck to determine the set of **universal constants**

$$\hbar = h/2\pi = 1.05 \times 10^{-27} \ \text{erg} \cdot \text{s} \tag{7.7}$$

$$k_B = 1.38 \times 10^{-16} \ \text{erg/K} \tag{7.8}$$

$$c = 3 \times 10^{10} \ \text{cm/s} \tag{7.9}$$

These constants, together with Newton's gravitational constant,

$$G = 6.67 \times 10^{-8} \ \text{cm} \cdot \text{g}^{-1} \cdot \text{s}^{-2},$$

allowed Planck to get a set of units[7] for mass length, time (and temperature) which are independent of specific bodies and substances and necessarily keep their meanings for all times and for all cultures... and can be designated as "natural units".

REFERENCES

[1] Julio A. Gonzalo (Coordinator) "Planck's Constant: 1900-2000" (Servicio de Publicaciones: Universidad Autónoma de Madrid, 2000) pp. 109-110.

[2] Max Born, Physik im Wandel meiner Zeit (4th ed.: Braunsweig: F. Vieweg, 1966) p. 244.

[3] Max Planck, "Dem Andenken and W.Wien "in K. Wien (ed) Wilhelm Wien: Aus dem Leben und Wirken eines Physikers (Leipzig: Johan A. Barth, 1930) pp. 139-140.

[4] Abhanlungen, vol. 2, p. 285.

[5] Abhandlungen, vol. 3, p. 64.

[6] R. Loudon, "The Quantum Theory of Light", 2nd. Ed., Chap. 1(Charedon Press: Oxford, 1983)

[7] Max Planck, "Wissenschflitche Selbsbiographie" (1948) p. 374.

1927: Uncertainty Principle / Primeval Atom

8. PLANCK'S UNITS AND HEISENBERG-LEMAITRE UNITS

Let us consider[1] a massive particle with a mass given by m_{Pl}, and a radius l_{Pl} such that its gravitational self-energy equals its relativistic energy as given by Einstein's relation:

$$m_{Pl}c^2 = \frac{Gm_{Pl}^2}{l_{Pl}}, \qquad (8.1)$$

Heisenberg's uncertainty principle (assuming $\Delta p \cong p$ and $\Delta x \cong x$) applied to this particle leads to

$$m_{Pl}c \cdot l_{Pl} \cong \hbar \qquad (8.2)$$

and combining Eqs. (8.1) and (8.2) we get

$$(m_{Pl} c \cdot l_{Pl}) \cdot c \cong \hbar c \cong G m_{Pl}^2, \tag{8.3}$$

i.e.

$$m_{Pl} \cong (\hbar c/G)^{\frac{1}{2}} \cong 2.17 \times 10^{-5} \, \text{g} \tag{8.4}$$

Planck's mass is therefore much larger than a baryon (proton, neutron) mass, $m_b \cong 1.67 \times 10^{-24}$ g .

Other physical quantities of Planck's particle can be obtained directly

$$l_{Pl} \cong \hbar/m_{Pl} c \cong (\hbar G/c^3)^{1/2} = 1.61 \times 10^{-33} \, \text{cm} \tag{8.5}$$

$$t_{Pl} \cong l_{Pl} / c \cong (\hbar G/c^5)^{1/2} = 5.36 \times 10^{-44} \, \text{s} \tag{8.6}$$

$$T_{Pl} \cong m_{Pl} c^2 / 2.8 k_B = (\hbar c^3/G)^{1/2} / 2.8 k_B \\ = 5.05 \times 10^{31} \, \text{K} \tag{8.7}$$

All these quantities are given in terms of the fundamental physical constants $\hbar$ (Planck quantum of action), G (Newton's gravitational constant), c (velocity of light in vacuum), and k_B (Boltzmann's constant). Planck's called them "natural units" and we can check that for our universe, which for Einstein and Lemaitre was **finite** (and could not be otherwise) they give surprisingly concordant numbers as shown in **Table I**. It will be seen in Chap. 11 that the presently observed Hubble's ratio (time dependent) $H_0 \cong 69$ km/s/Mpc and the present time (the time elapsed since the Big Bang to today) $t_0 = 13.8 \times 10^9$ years, allows one to use the solutions of Einstein's cosmological equations to get numbers for the total mass of the universe and for its characteristic (Schwarzschild) radius, and characteristic time

$$M_u = 0.7 \times 10^{55} \text{ g}, \quad R_+(t_+) = GM/c^2 = 5.3 \times 10^{26} \text{ cm}$$
$$t_+ = 600 \text{ Myrs} = 1.88 \times 10^{16} \text{ s}$$

The present radius and present time are
$R_0 \cong 1.5 \times 10^{28}$ cm, $\quad t_0 \cong 13.8$ Gyrs

Table I

Cosmic quantity	Planck Natural unit	Ratio
$M_u \cong 0{,}70 \times 10^{55}$ g	$m_{Pl} = 2.17 \times 10^{-5}$ g	3.22×10^{59}
$R_+ \cong 5.31 \times 10^{26}$ cm	$l_{Pl} = 1.61 \times 10^{-33}$ cm	3.29×10^{59}
$t_+ = 1.88 \times 10^{16}$ s	$t_{Pl} = 5.36 \times 10^{-44}$ s	3.52×10^{59}

Alternatively, taking the total cosmic mass M_u as the starting point, we can construct a set of units, which we will call Heisenberg-Lemaitre units, as follows

$$M_u = c^2 R_+ / G = 0.70 \times 10^{55} \text{ g} \tag{8.8}$$

$$l_{HL} = c \cdot \left(Mc^2 / \hbar \right) = 1.40 \times 10^{-112} \text{ cm} \tag{8.9}$$

$$t_{HL} = M_u c^2 / \hbar = 0.46 \times 10^{-102} \text{ s} \tag{8.10}$$

$$T_{HL} = M_u c^2 / 2.8 k_B = 1.63 \times 10^{91} \text{ K} \tag{8.11}$$

Eq. (8.10) sets a minimum time beyond which it becomes meaningless to speculate about cosmic dynamics. At lower times, the **uncertainty principle**

$$\Delta M_u c^2 \cdot \Delta t_{HL} > \hbar \tag{8.12}$$

forbids further speculation. In a sense, this takes care of the singularity at $R = 0$, $t = 0$. Heisenberg-Lemaitre's time $t_{HL} = 0.46 \times 10^{-102}$ s is many orders of magnitude smaller than Planck's time $t_{Pl} = 5.36 \times 10^{-44}$ s.

It may be noted[2] that the solution of Einstein's cosmological equation for an expanding universe implies at very early time that cosmic radius $R(t)$ grows with time as

$$R(t) = \text{Const.} \times t^{2/3} \quad \left(\text{Const} = R_+ \left[|k|^{1/2} c / R_+ \right]^{2/3} \right) \tag{8.13}$$

This means that, according to Einstein's compact solutions for an expanding universe, the growth factor between $t = t_{HL} = 0.46 \times 10^{-102}$ s and $t = t_{\text{infl}} \cong 10^{-35}$ s, as given by Alan Guth[3], is of the order of

$$R(t_{\text{infl}}) / R(t_{HL}) \cong (t_{\text{infl}} / t_{HL})^{2/3} = 7.8 \times 10^{44} \tag{8.14}$$

which is of the order of the inflationary growth factor $\left(\approx 10^{40} \right)$ assumed in Inflationary Cosmology to take place almost instantaneously at $t = t_{\text{infl}}$ in a singular cosmic phase transition. Eq. (8.13) gives a growth factor of the same order taking place continuously between 10^{-102} and 10^{-35} s which, it should be admitted, is not very different from "instantaneously". The growth takes place smoothly according to the compact solution of Einstein's cosmological equation. Therefore, no cosmic first order phase transition is needed.

Table II compares the Compton radius ($r_C = \hbar/mc$) and the Schwarzschild radius ($r_{Sch} = Gm/c^2$) for massive objects from the whole universe to the almost massless neutrino.

For a mass $m_{Pl} = (\hbar c/G)^{1/2} = 2.17 \times 10^{-5}$ g, the Compton radius $r_{Compton} = \hbar/m_{Pl} c = 1.61 \times 10^{-33}$ cm is equal to its Schwarzschild radius $r_{Sch} = Gm/c^2 = 1.61 \times 10^{-33}$ cm.

In the 1950's the **"Steady State Theory"**[4] of Gold, Bondi and Hoyle postulated "continuous creation of energy out of nothing" in order to keep constant cosmic density throughout the expansion for a universe in continuous growth. For years, the **"Steady State Theory"** was a rival of the **"Big Bang Theory"** originated in the **"Primeval Atom Theory"** of Lemaitre and then reformulated by Gamov, Alpher and Herman[5]. The discovery of the Cosmic Background Radiation[6] by Penzias and Wilson dealt a serious blow to the "Steady State Theory" which was unable to handle that cosmic radiation.

In 1993 at the time of the El Escorial Summer Course on Astrophysical Cosmology[7], t_0, the age of the universe (time elapsed since the Big Bang) was located somewhere between ten and twenty billion years. And the numerical value of Hubble's ratio $H_0 = \dot{R}_0/R_0$ somewhere between 50 and 100 km/s/Mpc. The interaction there with Ralph Alpher, John Mather, George Smoot, Stanley Jaki and others at the conference and after was extremely useful to accurately anticipate both quantities as $t_0 = 13.7 \times 10^9$ yrs and $H_0 = 65$ km/s/Mpc, in "Acta Cosmologica" (Cracow) in 1998, and it was reprinted five years later as an Appendix[8] in "Inflationary Cosmology Revisited".

Table II

Compton and Schwarzschild radii for massive objects

Object	m(g)	$r_{Compton}$ (cm) $(\hbar/mc)$	r_{Sch} (cm) (Gm/c^2)
Universe	0.70×10^{55}	5×10^{-93}	5.2×10^{26}
Galaxy	10^{44}	3.5×10^{-82}	7.4×10^{15}
Star	10^{33}	3.5×10^{-71}	7.4×10^{4}
Earth	6×10^{24}	5.8×10^{-63}	4.4×10^{-4}
Planck unit	2.1×10^{-5}	$\mathbf{1.61\times10^{-33}}$	$\mathbf{1.61\times10^{-33}}$
Baryon	1.67×10^{-24}	2.1×10^{-14}	1.2×10^{-52}
Electron	9.1×10^{-28}	3.8×10^{-11}	6.7×10^{-56}
Neutrino	2.2×10^{-35}	1.5×10^{-3}	1.6×10^{-63}

At El Escorial, I asked George Smoot (after his talk on "COBE's observations of the Early Universe") what did he think about the formal equivalence of the equations in the **Inflationary Theory** describing **sudden** cosmic expansion at constant density with the equations in the old **"Steady State Theory"** describing **continuous** cosmic expansion at constant density. I pointed out that it looked to me as if in the "Inflationary Theory" the same non energy conserving process was simply moved back in time, well beyond the reach of any possible observation. He answered, if I remember correctly, that in his opinion the two processes were not exactly identical, without denying certain similarities.

1927 was the year in which Werner Heisenberg published[9] his seminal work on the uncertainty principle. It was also the year in which Georges Lemaitre published[10] his original work on the solutions of Einstein's cosmological equation for a finite universe which opened the way for the Big Bang model.

REFERENCES

[1]Julio A. Gonzalo (Coordinator): "Planck's constants: 1900 – 2000" (Servicio de Publicaciones: Universidad Autónoma de Madrid, 2000) p. 103

[2]Julio A. Gonzalo, "Inflationary Cosmology Revisited" (World Scientific: Singapore 2005)

[3] Alan Guth, "The Inflationary Universe" (Perseus Books: Cambridge, Massachussetts, 1997)

[4]See f.i. H. Bondi, "Cosmology" (Cambridge University Press: Cambridge, 1952)

[5]See f.i. R.A. Alpher and R, Herman, "Physics Today" 41 (Part 1) 1988

[6]A.A. Penzias and R. Wilson, Astrophysical J. 142, 419 (1965)

[7] Julio A. Gonzalo, Jose L. Sánchez Gómez, and Miguel A. Alario (eds), "Cosmología Astrofísica" (Alianza Universidad: Madrid, 1996)

[8]Noé Cereceda, Ginés Lifante and Julio A. Gonzalo, "Acta Cosmologica" (Cracow) Fascicukus XXIV-2 (1998)

[9]Werner Heisenberg, "Zeitschrift für Physik" 43 (3-4): 172-198 (1927)

[10]Georges Lemaitre, "Annales de la Société de Bruxelles" 47, 49 (April 1927)

9. IMPLICATIONS OF A FINITE UNIVERSE

Today's theoretical cosmologists, even if they are conscious of the momentous implications of the universe being finite or infinite, tend to leave the question unanswered in the background. A few years ago David Spergel, NASA's theoretician in the WMAP team of which Ch. Bennet was the principal investigator, gave a very well attended- conference in Madrid on the latest developments in Cosmology. In it he said that, after being for a short time convinced that WMAP's data could be taken as a proof that our universe was **finite**, he had come to the conclusion that the proof was not really conclusive.

I put him an e-mail shortly afterward pointing out[1] that both Einstein and Lemaitre had assumed that the universe was finite and he wrote me back saying that not only Einstein and Lemaitre, but many other lesser figures were in favor of a finite universe. Still, he told me, it is very difficult to distinguish between a very large, enormous universe and an infinite one.

* * *

Of course, the controversy about the universe's finite or infinite character is nothing new. Even in classical antiquity Greek and Roman philosophers disagreed on the subject, with Platonists, Aristotelian, Stoic and Epicurean philosopher'sholding different views.

According to Stanley L. Jaki[2], **Newton** did not say anything about the universe in the first edition of the **"Principia"**. Richard Bentley in 1692, about ten years after the publication of the first edition, asked Newton's comments on what later would be known as the gravitational paradox of an infinite universe: in such universe the gravitational pull on any celestial body should be the same and equal to infinity in every direction. Newton had stated in an essay written in 1672 that the universe was finite in an infinite space, but before Bentley's question he was reluctant to come out openly on behalf of a finite or an infinite universe. It is well known that **Kant**, in his **"Critique of Pure Reason"** (1770) insisted that the universe is an unreliable notion: "the bastard product of the metaphysical cravings of the intellect". According to the Konigsberg philosopher, an amateur "scientist", science could not establish whether the universe was finite or infinite, atomistic or continuous. He was certainly not very convincing when he said: "If the world is a whole existing in itself, it is either finite or infinite. But both alternatives are false (as shown in the proofs of the antithesis and thesis respectively). It is therefore also false that the world (the sum of appearances) is a whole in itself." On the other hand, **Voltaire** in his book on the elements of Newtonian philosophy (available in German translation since 1741) and **Euler** (a great mathematician and scientist) had stated firmly the finitude of the universe, while **J.H.Lambert** (another great contemporary physicist and mathematician), in his **"Cosmological Letters on the Arrangement of the World Edifice"**, had pointed out "the impossibility of an actually realized infinite quantity", with which he endorsed unambiguously the finitude of the universe. In spite of all these competent opinions, Kant's rejection of a finite universe was highly influential.

In 1906, George Sorel, at the meeting of the Societé Francaise de la Philosophie made a remark about the restoration of the notion of a finite universe. His introductory question was: "Does one have the right to speak of the universe, or the ensemble of (all) things?".

Already **Gauss**, the prince of mathematicians, had noted, having probably in mind non-Euclidian geometries, that Kant's dicta on categories were sheer trivialities[3]. Then Zöllner, in Leipzig, was the first to work out a suggestion of **Riemann** dealing with finite matter in an endless space in 1872. **Szchwarschild** would work out in 1906 the space-time curvature of a finite observable universe[4].

In his memoir on the cosmological consequences of General Relativity, **Einstein** conveys the message that the universe is real and that it is no less specific than any real thing. This is most evident in his fifth memoir, where he gives formulae for the total mass and radius of the universe. He did not ask then however the most metaphysical question about the universe: Why is the universe what it is and not something else?

* * *

After this sketchy perspective on the question of the infinite or finite character of the universe, let us review briefly **Olber's paradox**, the most noteworthy scientific objection to a finite universe. Olbers paradox[5] is the riddle of the darkness of the night sky for those who take for granted the infinity of the universe: if there were infinite stars, in infinite galaxies, in an infinite universe, the line of sight of an observer should end always at the surface of an star, and then the whole night sky in a clear day should look bright, not dark.

William Olbers (1758-1840) did reformulate in 1823 the dark night paradox trying to save the then common presumption that the universe was infinite. Stanley L. Jaki makes clear in his carefully researched book on the subject that an infinite universe could readily pass for an ultimate entity, Nature, written with a capital letter, and therefore for a substitute of God. Jaki's book reprints essays by E. Halley (contemporary of Newton) J.P. Loÿs de Cheseaux and Olbers, and shows that the paradox could well have been named after Halley, who was the first to set it forth in Newton's time. More than a century after Olbers, H. Bondi, one of the original proponents of the "Steady State Theory" for the universe, in his Joule Memorial Lecture says the following: *"We can put a precise date to the moment when it (cosmology) became a scientific subject and left the realm of philosophical speculation. The date is 1826. In 1826 the German astronomer Olbers published a little investigation which, although I doubt whether he realized it made cosmology a science".* Olbers gave some reasons, not very convincing to explain why the night sky was dark, but the solution would come many years later with Einstein's General Theory of Relativity[6]. According to Einstein, the total mass of the universe can only be finite if scientific considerations about the universe as a whole are to remain meaningful. He says

"The total mass M of the universe, according to our views, is finite, and is in fact

$$M = \rho \cdot 2\pi^2 R^3 = 4\pi^2 \frac{R}{k} = \pi^2 \left(\frac{32}{k^3 \rho} \right)^{1/2} \qquad (9.1)$$

Thus the theoretical view of the actual universe, if it is in correspondence with our reasoning, is the following. The curvature of space is variable in time and place, according

> *to the distribution of matter, but we may roughly approximate to it by means of a spherical space".*

The important point here is that Einstein takes M to be finite to avoid physical inconsistencies. We know today that there are about 10^{11} to 10^{12} galaxies in the observable universe and about 10^{11} to 10^{12} stars of average mass, of the order of the Sun's mass, so that the total mass of the universe is on the order of $M_u \cong 10^{54}$ g.

In fact, the observable universe as we see it is bounded by the sphere of the CBR (Cosmic Background Radiation) whose minute anisotropies were detected by the COBE satellite and then, with increased precision, by the WMAP in 2003. The invisible microwave cosmic radiation, with a characteristic temperature of 2.726 K, fills the night sky, but its intensity is negligible in comparison with the light coming from visible galaxies made up of stars like our Sun, emitting visible light with a characteristic temperature of the order of 6300 K. In other words, the night sky is filled with invisible very week radiation, but this does not make the universe finite. On the contrary, beyond the CBR spherical surface (the surface of "last scattering") there is a spherical layer of finite "plasma universe whose matter and radiation mass can be obtained from observable cosmic quantities such as Hubble's ratio,

$$H_0 = \dot{R}_0 / R_0 \cong 69 \text{ km/s/Mpc} ,$$

the time elapsed since the Big Bang

$t_0 \cong 13.8 \times 10^9$ yrs, and the present CBR temperature $T_0 = 2.726$ K. Beyond that layer there is no space-time, no matter and no radiation.

Fig. 9.1 depicts the temperature evolution of the universe from the Heisenberg-Lemaitre temperature ($T \cong 10^{91}$ K) to the present ($T \cong 2.726$ K) as it expands and cools.

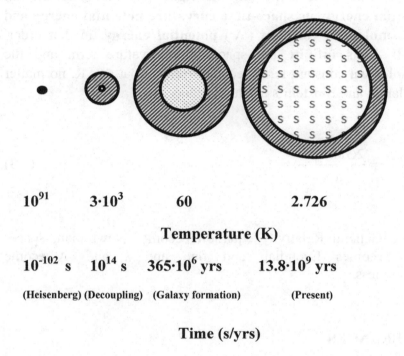

| 10^{91} | $3{\cdot}10^3$ | 60 | 2.726 |

Temperature (K)

| 10^{-102} s | 10^{14} s | $365{\cdot}10^6$ yrs | $13.8{\cdot}10^9$ yrs |

(Heisenberg) (Decoupling) (Galaxy formation) (Present)

Time (s/yrs)

Fig. 9.1 Temperature / Time evolution of the universe from Heisenberg's time to present.

It is instructive to see directly the implications of an infinite mass for the universe in Einstein's **cosmological equation**

$$\frac{1}{2} m_G \dot{R}^2 = \frac{2GMm_G}{R} - kc^2 m_G + \frac{\Lambda}{3}c^2 R^2 m_G \qquad (9.2)$$

where m_G is the mass of a galaxy moving away at a distance R from the singularity ($t = 0$, $R = 0$) at as speed $\dot{R}$, and M is the total mass of the universe. The single term of the left gives its **kinetic energy**. The three terms on the right give its **gravitational potential energy**, its **space-time curvature potential energy** and its **cosmological constant** (Λ) **potential energy**, in that order. But if M is **infinite** the space-time curvature term and the cosmological constant term become irrelevant at any R, no matter how large, and Einstein's equation reduces to

$$\frac{1}{2}m_G\dot{R}^2 = \frac{2GMm_G}{R} \to \infty \qquad (9.3)$$

The General Relativity equation becomes Newtonian, space-time becomes Euclidian and flat, and k and Λ become meaningless.

REFERENCES

[1]See f.i. Julio A. Gonzalo "Cosmic Paradoxes" (World Scientific: Singapure, 2012)

[2]Stanley L. Jaki, "God and the cosmologists" (Real View Books: Fraser, Michigan, 1998. Revised and entirely reset 2nd ed. With a Postcript, 1998.(First published in 1898).

[3]As Gauss wrote on Nov. 1, 1844, to H.C. Schumacher, Kant's distinction "between analytic and synthetic propositions is one of those things that either run out on triviality or are false". "Gauss Gesammelte Werke" (Göttingen: K. Gesellschaft der Wissenschaften, 1870-1933), vol. 12, p. 63.

[4]See Stanley L. Jaki, "The Milky Way": An Elusive Road for Science" p. 276 (Science History Publications: New York, 1972).

[5]Stanley L, Jaki, "The Paradox of Olbers Paradosx" (Real View Books: Pickney MI, 2000)

[6]A. Einstein in "The Principle of Relativity: A collection of Original Memoirs on the Special and the General Theory of Relativity, eds. H.A. Lorentz, A. Einstein, H. Perret and G.B. Jeffrey (Dower: New York, 1952) pp 110-188.

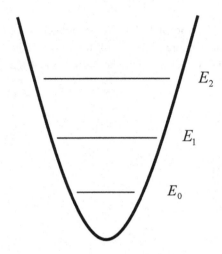

10. COSMIC ZERO-POINT ENERGY

The **zero point energy**[1] for electromagnetic radiation confined in a finite expanding spherical cavity of radius R, no matter how large, is given by

$$\Delta \overline{E}_0(R) = \sum_k \frac{1}{2} h \nu_{\bar{k}} \cong \int_{\nu_{\min}(R)}^{\nu_{\max}(R_{\min})} V \cdot \frac{h}{c^3} \nu_{\bar{k}}^3 d\nu_{\bar{k}} \qquad (10.1)$$

where $V \cong L^3 \cong \frac{1}{2}(2R)^3$, and $\nu_k = \frac{c}{R}$, $d\nu_k = \frac{c}{R^2} dR$.

Therefore taking $R_{\min} = R_H = h/Mc$, which is Heisenberg's radius (or Compton radius) for a universe of mass M, enormous but finite, we have (except for a factor of order unity)

$$\Delta \overline{E}_0(R) \cong -ch \int_R^{R_H} \frac{dR}{R^2} = -ch \left[-\frac{1}{R} \right]_R^{R_H} = ch \left[\frac{1}{R_H} - \frac{1}{R} \right] \qquad (10.2)$$

and taking into account that $ch/R_H = Mc^2$, for $R \gg R_H$ the zero point energy would be given by

$$\Delta \overline{E}_0(R) \cong \frac{ch}{R_H} \left[1 - \frac{R_H}{R} \right] \cong Mc^2 \quad (R \gg R_H) \qquad (10.3)$$

In order to see how $\overline{E}_0(R)$ evolves with time from $t = t_H = 0.46 \times 10^{-102}$ s, to "decoupling" (atom formation), $t_{dec} \cong 10^{14}$ s, i.e. during the **plasma phase** of cosmic expansion at which ρ_r (radiation) $= \rho_m$ (matter), as discussed below, and then from $t_{dec} \cong 10^{13}$ s. to the present epoch $t_0 = 13.8 \times 10^9$ yrs, i.e. during the **transparent phase** of the expansion, at which ρ_r decreases continuously from $\rho_{rdec} = \rho_{mdec}$ to its present value $\rho_{ro} \ll \rho_{mo}$, it is necessary to specify the **equation of state**[2,3], $R(T)$, for the **plasma phase** and for the **transparent phase** in which, at some time stars and galaxies begin to form to become fully formed at a latter time.

In the **plasma phase**, after massive charged particles and electrons are formed, radiation and particles move in unison away from the center of the sphere specified by t_H (almost indistinguishable from the vanishing sphere specified by $t = 0$ the Big Bang singularity). In due time material particles protons ^{4}He nuclei, electrons, scatter photons (radiation) and photons push away charged particles resulting in cosmic expansion.

Before decoupling (atom formation) and at least since primordial nucleosynthesis (^{4}He formation), i.e. since $T_{ns} \cong 4.6 \times 10^8$ K,

$$\rho_r(T) = \sigma T^4 = \rho_m = Mc^2 \bigg/ \frac{4\pi}{3} R^3, \tag{10.4}$$

which gives the **plasma phase equation of state**

$$R^3 T^4 = Mc^2 \bigg/ \frac{4\pi}{3} \sigma = \text{const}, \tag{10.5}$$

leading to

$$\Delta \overline{E}_0 = \overline{E}_0(R) - \overline{E}_0(R_H) = Mc^2 \left[1 - \frac{R_H}{R} \right] =$$

$$= Mc^2 \left[1 - \left(\frac{T}{T_H} \right)^{4/3} \right] \cong Mc^2 \tag{10.6}$$

On the other hand, in the *transparent phase*, after decoupling (atom formation), $T_{af} \cong 3000\,\text{K}$, neutral atoms have been formed, violent scattering of photons has practically ceased, and the total number of photons (as the total number of baryons tied up in atoms, mainly H and ^{4}He nuclei at the beginning) becomes fixed in the universe. Then,

$$n_\gamma(T) = \frac{\sigma T^4}{2.8 k_B T}; \quad n_m = \frac{Mc^2 \bigg/ \dfrac{4\pi}{3} R^3}{m_b} \tag{10.7}$$

$$\left(n_\gamma / n_m = \text{const at } t > t_{dec} \right)$$

which results in the **cosmic equation of state at the present phase**

$$R^3 T^3 = \text{const},\tag{10.8}$$

resulting in $\rho_\gamma = (\rho_\gamma)_{dec} (T/T_{dec})$, and, therefore

$$\Delta \bar{E}_0 = \bar{E}_0(R) - \bar{E}_0(R_{dec}) = Mc^2 \left[1 - \frac{T}{T_{dec}} \right] < Mc^2 \tag{10.9}$$

Eq. (10.8) means that since T_{dec} to T_o the radiation energy density decreases with respect to the matter energy density as $\rho_r / \rho_m \cong T/T_{dec}$.

Since after decoupling (atom formation)

$$\rho_\gamma = \rho_r (T_{dec}) \left(\frac{T}{T_{dec}} \right)^4$$

$$\rho_m = \rho_m (R_{dec}) \left(\frac{R_{dec}}{R} \right)^3$$

and $RT = R_{dec} T_{dec} = \text{const}$,

$$\frac{\rho_r}{\rho_m} = \left(\frac{\rho_r}{\rho_m} \right)_{dec} \left(\frac{T}{T_{dec}} \right) = \frac{T}{T_{dec}} \tag{10.10}$$

and, also, since denoting by ρ_{zp} the "zero point" energy density

JULIO A. GONZALO

$$\rho_{zp} = \rho_r (T_{dec}) \left(1 - \frac{T}{T_{dec}} \right)$$

$$\rho_m = \rho_m (R_{dec}) \left(\frac{R_{dec}}{R} \right)^3$$

and again $RT = R_{dec} T_{dec} = \text{const}$,

$$\frac{\rho_{zp}}{\rho_m} = \left(\frac{\rho_r}{\rho_m} \right)_{dec} \left(1 - \frac{T}{T_{dec}} \right) = 1 - \frac{T}{T_{dec}} \tag{10.11}$$

and putting together Eqs. (10.10) and (10.11) we get

$$\frac{\rho_r}{\rho_m} + \frac{\rho_{zp}}{\rho_m} = \left[\frac{T}{T_{dec}} + \left(1 - \frac{T}{T_{dec}} \right) \right] = 1$$

which insures energy conservation *taking into account* the temperature dependence of the *finite cosmic zero-point energy* throughout the expansion. In other words,

$$\Delta\rho_\gamma + \Delta\rho_{zp} = \Delta\rho_m, \text{ for all } T. \tag{10.12}$$

REFERENCES

[1] R. Loudon "The Quantum Theory of Light", 2nd ed. (Charendon Press: Oxford 1983) p 139.

[2] N. Cereceda, M.I. Marqués, G. Lifante and J.A. Gonzalo, in "Frontiers of Fundamental Physics", B.G. Sidarth, A.A. Faus and M.J. Fullana eds, Universidasd Politécnica, Madrid, 2006, AIP Conf. Proc 905, (2007) pp 6-12

[3] Julio A. Gonzalo, "The Intelligible Universe", 2nd ed., pp. 312-317 (World Scientific: Singapore, 2008).

$$\dot{R}^2 = \frac{2GM}{R} - kc^2 + \frac{\Lambda}{R^2}c^2$$

11. RIGOROUS SOLUTIONS OF EINSTEIN'S COSMOLOGICAL EQUATION

Manuel Alfonseca and myself submitted recently to arXivs.org our work **"Comment on the 1% Concordance Hubble Constant"**. In it a summary of the solutions of Einstein cosmological equation for an **Open Friedman-Lemaitre universe** and a flat **Lambda Cold Dark Matter universe** model are examined.

C.L. Bennet et al. have recently observed that the accurate determination of the Hubble constant $(\dot{R}_0/R_0)$ for $z \cong 0$ has been, and is, a central goal in observational astrophysics. After a careful analysis of the existing data they conclude that $H_0 = 69.6 \pm 0.7 \, \mathrm{kms^{-1}Mpc^{-1}}$. This is almost compatible with $H_0 = 73.8 \pm$

$2.4 \, \text{kms}^{-1}\text{Mpc}^{-1}$ as recently evaluated by Riess et al. In this comment we note that the difference is significant at least in one respect: assuming $t_0 = 13.7$ Gyrs ($\pm$ 0.5 %) the value given by C. L. Bennet et al. results in $H_0 \, t_0 = 0.975 < 1$, which is compatible at least marginally with an open universe ($k < 0$, $\Lambda = 0$), while the value given by Reiss et al, $H_0 \, t_0 = 1.034 > 1$, is not. NASA's James Webb Space Telescope, to be launched in 2017, may be expected to determine H_0 with an accuracy better than 1%.

It is well known that Hubble's original attempt (1929) to make a quantitative evaluation of the velocity – distance ratio for distant galaxies involved serious systematic errors. For a long time afterwards, Sandage, Hubble's successor at Mount Wilson Telescope favored a Hubble constant value of $H_0 \cong 50$ km s^{-1}Mpc^{-1}, while Vaucouleurs, another distinguished astronomer, favored $H_0 \cong 100$ km s^{-1}Mpc^{-1}. NASA's Hubble Space Telescope produced $H_0 \cong 72 \pm 8$ km s^{-1}Mpc^{-1} (Freedman et al. 2001)[1]. More recently (Riess 2014)[2] have given $H_0 \cong 73.8 \pm 2.4$ km s^{-1}Mpc^{-1}, which in principle reduces the uncertainties to 3%. The problem is by no means settled and the quoted uncertainties may be a little optimistic.

Bennet et al.[3] have published a careful re-examination of the self-consistency of H_0 measurements as given by the WMAP and Planck satellites and some ground based telescopes. The best fit, obtained after a meticulous evaluation, gives $H_0 \cong 69.6 \pm 0.7$ km s^{-1}Mpc^{-1}. Hopefully NASAs James Webb Space Telescope, to be launched in 2017, will give a vastly improved value for H_0, both for the central value and for the reduced uncertainty.

On the other hand, as pointed out long ago (Beatriz Tinsley, 1977)[4] the dimensionless product $H_0 t_0$ can be especially advantageous to characterize quantitatively the solutions of Einstein's cosmological equations. Using the analytic solutions of Einstein's equations for $\Lambda > 0$ (Lambda Cold Dark Matter model with k = 0) and for $k > 0$ (Open Friedmann-Lemaitre model with $\Lambda = 0$) it is easy to check (Gonzalo and Alfonseca, 2013) that

$$\frac{2}{3} \leq H_0 t_0 \leq \infty \qquad \text{(LCDM model)}, \qquad (11.1)$$

and

$$\frac{2}{3} \leq H_0 t_0 \leq 1 \qquad \text{(OFL model)}, \qquad (11.2)$$

Here the numerical value depends principally on H_0, since t_0 was determined with magnificent accuracy by WMAP in 2003 and was confirmed by **Planck Satellite** in 2013 as $t_0 = 13.7 \pm 0.1$ Gyrs.

It may be instructive to take a look at characteristic cosmological parameters such as $\Omega = \Omega_m + \Omega_r$ (matter mass density plus radiation mass density divided by the critical density) or z_{Sch} (the hypothetical maximum redshift corresponding to $R = R_{Sch}$ for the universe, obviously greater than $(z_{obs})_{max}$ for the most distant galaxies (or quasars) observable) as a function of H_0 for the analytic solutions of Einstein's cosmological equations (Gonzalo and Alfonseca, 2013) in three cases: (a) a **flat** LCDM (Lambda Cold Dark Matter) universe with $\Lambda > 0$; (b) an **open** OFL (Open Friedman Lemaitre) universe with $k < 0$; and (c) a **"mixed"** ($\Lambda < 0$, $k < 0$) universe in a wide interval for H_0 encompassing the interval $65 \leq H_0 \leq 75$.

For a **flat universe** the analytic solutions of Einstein's cosmological equations ($\Lambda < 0$, $k < 0$) are given (Gonzalo and Alfonseca, 2013)[5] by

$$t(y_L) = \frac{2}{3}\left(\frac{\Lambda}{3}c^2\right)^{-1/2} y_L = \frac{2R_L}{3c}\left(\frac{R_L c^2}{2GM_L}\right)^{1/2} y_L,$$

$$R(y_L) = R_L \sinh^{2/3} y_L \qquad (11.3)$$

where $\Lambda > 0$ is Einstein's cosmological constant, c is the velocity of light, G is Newton's gravitational constant, M_L the total mass of the universe,

$$R_L = [2\,G\,M_L\,/\,(1/3)\Lambda c^2]^{1/3}$$

the cosmic radius when the cosmic density parameter Ω equals $(1/2)$ and y_L an auxiliary parameter $y_L \equiv \sinh^{-1}(R/R_L)^{3/2}$ going from $y_k = 0$ at the singularity (Big Bang) to $y_k \to \infty$ in the very distant future.

Then, at present,

$$R_0 = c/H_0 = R_L \sinh^{2/3} y_{L0} \tag{11.4}$$

It is easy to check from Eqs. (11.3) that

$$H(y_L) = \left(\frac{\Lambda}{3}c^2\right)^{1/2} \frac{\cosh y_L}{\sinh y_L}$$

$$H(y_{L0}) = \left(\frac{\Lambda}{3}c^2\right)^{1/2} \frac{\cosh y_{L0}}{\sinh y_{L0}}, \tag{11.5}$$

$$H(y_L)t(y_L) = \frac{2}{3}\frac{y_L}{\tanh y_L}$$

$$H_0 t_0 = \frac{2}{3}\frac{y_{L0}}{\tanh y_{L0}} \text{ (dimensionless)}, \tag{11.6}$$

$$\Omega(y_L) = 1 - \tanh^2 y_L$$

$$\Omega_0 = 1 - \tanh^2 y_{L0} \text{ (dimensionless)} \tag{11.7}$$

In this model of the universe, as can be observed in Table I at the end of the paper, the candidate most distant galaxy (which has a red-shift of 10.8) sent its light towards us well before the universe reached its Schwartzschild radius, which means that we were inside a black hole. Of course, at that time we, or rather, the seeds of the solar system if any, were also inside the same hole. At that time, for this model, the density parameter of the universe was almost exactly 1, which means that the contribution of the cosmological constant (which nowadays is computed to contribute about 70%) was the negligible.

Notice that $\Omega = \frac{2}{3}$ corresponds to the inflection point between decelerated expansion and accelerated expansion in the **flat model** (the point when the acceleration was zero). This point does not exist in the **open model**.

For an **open universe** the analytic solutions of Einstein's cosmological equations ($\Lambda = 0$, $k < 0$) are likewise given by

$$t(y_k) = \frac{R_+}{c|k|^{1/2}} (\sinh y_k \cosh y_k - y_k) \ ,$$

$$R(y_k) = R_+ \sinh^2 y_k \qquad\qquad (11.8)$$

where $k < 0$ is the negative space time curvature (which might be associated with an antigravitational radiation pressure), and $R_+ = (2GM_k/|k|c^2)$ is the cosmic radius when $\Omega = \frac{1}{2}$, (in this case corresponding to the cosmic Schwarzschild radius)

M_k is the total mass of the universe and y_k is the corresponding auxiliary parameter going from zero to infinity.

At present,

$$R_0 = (c/H_0)/\tanh y_0 = R_+ \sinh^2 y_{k0} \qquad (11.9)$$

and then

$$H(y_k) = \frac{|k|^{1/2} c \cosh y_k}{\sinh^3 y_k}$$

$$H(y_{k0}) = \frac{|k|^{1/2} c \cosh y_{k0}}{\sinh^3 y_{k0}}, \qquad (11.10)$$

$$H(y_k)t(y_k) = \frac{1}{\tanh^2 y_k} - \frac{y_k}{\tanh y_k \sinh^2 y_k}$$

$$H_0 t_0 = \frac{1}{\tanh^2 y_{k0}} - \frac{y_{k0}}{\tanh y_{k0} \sinh^2 y_{k0}}, \qquad (11.11)$$

$$\Omega(y_k) = 1 - \tanh^2 y_k$$

$$\Omega_0 = 1 - \tanh^2 y_{k0} \qquad (11.12)$$

In this model of the universe, as can be observed at Table I, for $H_0 \geq 65$ the candidate most distant galaxy (which has a red-shift of 10.8) sent its light towards us after the universe reached its Schwarzschild radius (and the density parameter was equal to 1/2). At that red-shift, the density parameter was smaller than 2/3 for all H_0 values analyzed, which means that the contribution of the space-time curvature was above 1/3. With this model, it may be assumed that galaxy formation started after the universe ceased behaving as a black hole.

Finally, for a **"mixed" universe** ($\Lambda > 0$, $k < 0$) the interpolated solutions of Einstein's can be given by

$$\Omega_x = (1 - \tanh^2 y_L)^x (1 - \tanh^2 y_k)^{1-x}$$

In this model of the universe, as can be observed at Table I for x = 0.5, the density parameter is intermediate between those of the other two models, somewhat nearer to the open than to the flat for x = 0.5.

We have assumed that Λ = const > 0 for a flat universe (which leads to $\dot{R}(\infty) \to \infty$) and k = const < 0 for an open universe (which leads to $\dot{R}(\infty) \to |k|^{1/2}c$, i.e. $\dot{R}(\infty) \to c$ for k = -1).

Fig. 11a gives Ω_0 for an open (OFL) and a flat (LCDM) universe as a function of H_0 assuming $t_0 = 13.7$ Gyrs. H_0 as given by (Riess et at 2011) results in $H_0 t_0$ to the right of $H_0 t_0 = 1$ while H_0 as given by (Bennet et al (2014) results in $H_0 t_0$ to the left of $H_0 t_0 = 1$, as noted previously.

Fig. 11b gives z_{Sch}, the redshift corresponding to light emitted at R_{Sch} (the radius at which the universe ceased behaving as a black hole, which could correspond to the start of galaxy formation in the open model).

The data shown in Figs. 11a and 11b are summarized in Table I.

Note that we can approximate an average value of the density parameter <Ω>from the first galaxies to us, by computing the half sum of $\Omega(z_{1\ gal})$ to $\Omega(0)$, with both models, for $H_0 = 69.6$. The results are <Ω>$_{flat}$ = (1 + 0.2805)/2 = 0.6402 and <Ω>$_{open}$ = (0.5+0.0084)/2=0.254, respectively.

As it is known, NASA's JWS Telescope expected to be launched on schedule in 2017 may improve considerably the accuracy with which H_0 is known. No great surprises are generally expected but certainly, reliable new data for H_0 should be welcome.

Table I

Flat universe: $R_0 = c/H_0$, $T_0 = 2.726$ K, $R_0 T_0 = RT$, $R_{Sch} = 2GM_L/c^2$

y_0	0.98	1.05	1.12	1.18	1.25	1.31	1.36
H_0(km/sMpc)	62	64	66	68	70	72	74
$\Lambda \times 10^{21}$	6.86	7.9	8.91	10.0	11.0	12.1	13.2
Ω_0	0.43	0.39	0.35	0.31	0.28	0.25	0.23
$T_{Sch}(K)$	6.3	7	7.8	8.7	9.7	10.7	11.8
z_{Sch}	1.32	1.59	1.87	2.20	2.55	2.93	3.35

Open universe: $R_0 = (c/H_0)/\tanh y_0$, $T_0 = 2.726$ K, $R_0 T_0 = RT$, $R_{Sch} = 2GM_k/|k|c^2$

y_0	1.59	1.81	2.06	2.41	3.03	∞
H_0(km/sMpc)	62	64	66	68	70	71.374
Ω_0	0.15	0.10	0.062	0.032	0.009	0
$T_{Sch}(K)$	15.2	23.9	41.0	83.8	289	-
z_{Sch}	4.56	7.78	14.04	29.74	105	-

Numerical data: Fig 11(a) / (b)

$$t_0 = 13.8 \, \text{Gyr}$$
$$H_0 t_0 = 1 \leftrightarrow H_0 = 70.856$$
$$H_0 t_0 = 1.0303 \leftrightarrow H_0 = 73.0 \; (Riess \; et \; al.)$$
$$H_0 t_0 = 0.9823 \leftrightarrow H_0 = 69.6 \; (Bennet \; et \; al)$$
$$T_0 = 2.726 \, \text{K}$$
$$R_0 T_0 = RT$$

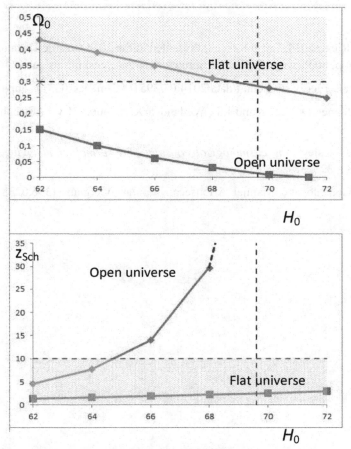

Fig. 1a: Ω_0 for an open (OFL) and a flat (LCDM) universe.
Fig.1b: gives z_{Sch}, the redshift corresponding to light emitted at R_{Sch}

REFERENCES

[1] L. Freedman, B. F. Madore, B. K. Gibson, L. Ferrarese, D. D. Kelson, S. Sakary, J. R. Monld, R. C. Hennicutt, H. C. Ford, J. A. Graham, J. P. Huchra, S. M. G. Hughes, G. D. Illingworth, L. M. Macri and P. B. Stetson, *Final Reaults form the Hubble Space Telescope Key Project to Meassure the Hubble Constant.* ApJ, 533: 47 – 72, 2001.

[2] A. G. Riess, 2014, Local Results, Presented in the 2014 Cosmic Distance Scale Workshop, http://www.stsci.edu/institute/conference/cosmic-distance/

http://realserver4v.stsci.edu/t/data/2014/03/3951/AdamRiess033114.mp4

[3] C.L. Bennet, D. Larson and J. L. Weiland, arXiv: 1406.171 v 1 (astroph. CO) 6 Jun 2014.

[4] B. M. Tinsley, *The cosmological constant and cosmological change*, Phys. Today 30, 32 – 38.

[5] J. A. Gonzalo and Manuel Alfonseca, http://arxiv.org/abs/1306.0238, 2 Jun 2013.

12. ON THE EVIDENCE FOR DARK MATTER, DARK ENERGY & ACCELERATED EXPANSION

Dark matter

One of the oldest arguments[1,2] to postulate large amounts of invisible dark matter is the non-Keplerian velocity rotation of stars in galaxies.

If the rotation velocity of stars in large spiral galaxies about its center were Keplerian $v(r)$ would decrease with r as $r^{-1/2}$ because

$$\frac{1}{2}mv^2 = \frac{GMm}{r} \tag{12.1}$$

and then

$$v = (2GM)^{1/2} r^{-1/2} \tag{12.2}$$

but $v(r)$ tends to be rather independent of r as r increases beyond a certain r^* up to a certain $r_m \gg r^*$. Therefore, goes the argument, galaxies must contain much more mass somewhere than it was originally assumed.

However, the argument is flawed. In a planetary (Keplerian) system, like our Solar System, almost all of the mass is concentrated in the center and then Eq. (12.2) applies. On the other hand, in a spiral galaxy (Fig. 12.1) the total mass is distributed in such a way that

$$M(r) = M^* \left(r/r^*\right) \text{ for } 0 < r < r^* \tag{12.3}$$

$$M(r) = M^* + \Delta M(r) \cong M^* + M^* \left(\frac{r}{r^*} - 1\right) \cong$$
$$\cong M^* \left(\frac{r}{r^*}\right) \text{ for } r^* < r < r_m \tag{12.4}$$

Therefore,

$$v(r) = \left(\frac{2GM^*}{r^*}\right)^{1/2} \left(\frac{r}{r^*}\right) \text{ for } 0 < r < r^* \tag{12.5}$$

$$v(r) = \left(\frac{2GM^*}{r}\right)^{1/2} \left(\frac{r}{r^*}\right)^{1/2} = \left(\frac{2GM^*}{r^*}\right)^{1/2} =$$
$$= v_m \text{ (constant) for } r^* < r < r_m \tag{12.6}$$

Galaxy: Side View

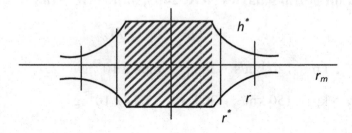

$$M^* = \left(\pi \cdot r^{*2}\right) \cdot h^* \rho^*$$

$$M_G = M^* + \Delta M = M^* + M^*\left(\frac{r_m}{r^*} - 1\right)$$

Galaxy: Top view

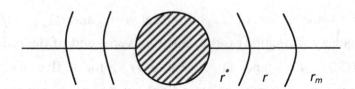

Fig. 12.1 Mass distribution in a spiral galaxy: (a) Side view; (b) Top view

Table 12.1 gives data[3] for two spiral galaxies (a) NGC 2403, and (b) NGC 3188 which allow the determination of their respective masses by means of

$$M_G = \left(\frac{v^{*2} r^*}{2G}\right)\left(\frac{r_m}{r^*}\right) \quad \text{where } v^* = v_m \text{ is } v(r)$$

(12.7)

for r in the interval $r_m^* \leq r \leq r_m$.

Table 12.1

Data for spiral galaxies NGC 2403, and NGC 3188

Galaxy	r^*	v^*	r_m	M_G
NGC 2403	5 kp	120 km/s	20 kp	6.64×10^{43} g
NGC 3188	5 kp	150 km/s	30 kp	15.54×10^{43} g

Taking into account that the mass of a typical star like the Sun is of the order of $M_S = 2 \cdot 10^{33}$ g, the first galaxy has a mass equivalent to 3.3×10^{11} solar masses and the second 7.7×10^{11} solar masses, reasonably large figures which do not include any dark matter.

It must be pointed out, as we have seen in Chap. 7, that the density parameter $\Omega_m = \rho_m / \rho_{mc}$ is strongly time dependent.

It is of the order of $\Omega_{mo} \cong 0.02$ at $t = t_0$ and $\Omega_{mgf} \cong 1/2$ for $t = t_{gf}$ (galaxy formation) for an open universe and of the order of $\Omega_{mo} \cong 0.30$ at $t = t_0$ and $\Omega_{mgf} \cong 1$ at $t = t_{gf}$, for a flat one. Our telescopes are seing now galactic light emitted relatively recently together with galactic light emitted billions of years ago, which can account for a lot of apparent dark matter.

Dark energy

Dividing Einstein's cosmological equation by the kinetic energy term at the right and side of the equality one has

$$1 = (\Omega_m + \Omega_r) + \Omega_k + \Omega_\Lambda, \quad \Omega = \Omega_m + \Omega_r, \tag{12.8}$$

where, at present, Ω_{mo} (matter density parameter) is much greater than Ω_{ro} (radiation density parameter) and therefore $\Omega_o \cong \Omega_{mo}$.

As we have noted Ω_k is associated to the space-time curvature potential energy of the universe if $k \neq 0$, and to Ω_Λ, the cosmological constant potential energy of the universe, if $\Lambda \neq 0$.

Since 1998, after the first reports[4,5] of cosmic accelerated expansion were circulated, most theoretical cosmologists decided that Ω_{mo} including dark matter could not be greater than 1/3, that Ω_k should be zero, and that, therefore, Ω_Λ should be of the order of 2/3, and begun calling it "**dark energy**".

As mentioned in the previous Chapter there is still considerable uncertainty, probably larger than **1%** in the present value of Hubble's ratio ($H_0 = \dot{R}_0 / R_0$). The observational evidence does not exclude $\Omega_k \neq 0$ in which case Ω_k could be called "dark energy" instead of Ω_Λ if one whishes so.

Accelerated expansion

As noted, in 1998 the first reports of an unexpected accelerated expansion deduced from galaxies up to redshifts slightly above $z \cong 1$ were immediately taken as a proof that the cosmological constant Λ, originally introduced by Einstein to produce a static universe (and later discarded) was $\Lambda > 0$. Sam Perlmutter, Adam Riess and Brian Schmidt, received the 2011 Physics Nobel Prize for the discovery of accelerated expansion. Perlmutter noted very early that an increase in cosmic dust for relatively distant galaxies might produce a decrease in their brightness (artificially enhancing their apparent magnitude *m*) and so it might contribute to enhance

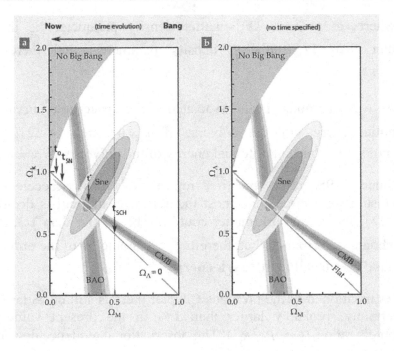

Figs. 12.2 (a) and (b) give, respectively, (a) Ω, Ω_k as a function of temperature/time/ y_k (cosmic parameter) in an open universe, and redshift; and (b) Ω, Ω_Λ as a function of temperature/time/ y_L (cosmic parameter in a **flat** universe) and redshift.

the estimated speed of the accelerated expansion. Incidentally one would have expected that new data for higher redshifts would have made the case stronger for accelerated expansion since 1998. In any case even if $m(z)$ is confirmed (as it seems) to **go up** with increasing z beyond $z = 1$, after taking into consideration the effect of cosmic dust the observational evidence is not yet fully satisfactory.

On the other hand, a more rigorous representation of distance $r(z)$ vs. velocity $\dot{r}(z)$ taking into account straightforward purely relativistic effects shows that $r(z)$ vs. $\dot{r}(z)$ **goes up** smoothly with z, resulting in an **"effective" accelerated expansion**, very small at z slightly above 1, but more noticeable at higher redshifts. (Fig 12.2). As shown[6] in **"Constraints on the general solutions of Einstein's cosmological equations by H_0t_0: a historical perspective"** (Julio A. Gonzalo and Manuel Alfonseca) $\log_{10}(r/R_0)$, proportional to the apparent magnitude of distant galaxies ($z \approx 1$ and beyond) as a function of $\log_{10}(v/c)$, proportional to the $\log_{10}$ of recession velocity is given by

$$r = R_0 \frac{z}{1+z} \rightarrow [r(z)]_{z<<1} \cong R_0 z \qquad (12.9)$$

$$v = c \frac{(1+z)^2 - 1}{(1+z)^2 + 1} \rightarrow [v(z)]_{z<<1} \cong cz \qquad (12.10)$$

which for $z<<1$ results in $\log_{10} r(z)$ as a function of $\log_{10} v(z)$ in perfect agreement with Hubble's law, but begins to show at $z \geq 1$ an **upwards trend**. Then $\log_{10}(r/R_0)$, proportional to the apparent magnitude, as a function of $\log_{10}(v/c)$, is no longer proportional to $\log_{10} z$, which can be taken as an accelerated expansion, but it has nothing to do with any non-zero cosmological constant Λ.

Therefore, even if the minor **upwards trend in magnitude vs. redshift** is confirmed more conclusively for $z > 1$, this would not mean necessarily that $\Lambda \neq 0$. Dark energy understood as associated with a non-zero Cosmological Constant cannot be inferred unambiguously on the grounds of the presently available observational evidence. In fact, one should wait for more conclusive data on $\Omega_m(z)$, and $H(z)$ to decide whether the universe is **flat** or **open**.

The evidence seems to be in favor of a universe **single** (no "multiverses"), **finite** (not infinite), and **open** (not flat). As always in physics, the observational data will have the last word.

REFERENCES

[1]See f.i. V.C. Rubin, W. K. Ford and N. Thonnard, Astrophys. J. 225, L 107; 238, 471 (1990).

[2]James T. Dwyer (jtdwyer @ rocketmail.com), Private Communication (2014)

[3]J. Peebles, "Principles of Physical Cosmology" p. 47 (Princeton U. Pres: Princeton, New Jersey, 1993).

[4]S. Perlmutter et al. Astrophys. J. 517, 565 (1998)

[5]A.G.Riess et al. Astron. J. 116, 1009. Also see B. Schmid et al Astrophys. J. 507, 46 (1998).

[6]Julio A. Gonzalo and Manuel Alfonseca, arXiv (2013)

13. ON PHYSICS AND PHILOSOPHY

In his book "Physics and Beyond"[1], Werner Heisenberg says that one of his main objectives is to reexamine the contradictory meanings given today to words such as God, democracy, freedom, justice, love, peace and brotherhood. In Chapter 7, entitled "Science and Religion", the author summarizes his reminiscences of conversations with a few young brilliant physicists, Schrodinger, Pauli, Dirac, present in 1927 at the Solvay Conference in Brussels. That conference funded by the Solvay Company, was organized with funds destined to reward whoever was capable of demonstrating the famous "Fermat Last Theorem". (There is no triplet made up of integers fulfilling the equality

$x^n + y^n = z^n$ for n integer beyond $n = 3$). No such proof of Fermat Last Theorem was forthcoming at the time, but the conference served as an exceptional forum for the young creators of Quantum Mechanics and its pioneers, Planck, Einstein, Bohr and de Broglie.

In those conversations, according to Heisenberg, the ideas of Planck and Einstein were hotly debated. The two greatest 20th century physicists saw a natural connection between their realistic world views and the western religious tradition, in contrast with Bohr. For Planck and Einstein, religion and science were not incompatible. According to Heisenberg, Planck did not see any contradiction, because, he said, science and religion looked at different aspects of reality. In Heisenberg's words Planck held that science had to do with the material, objective world, while religion had to do with the world of values. Our decisions are often guided by educational, environmental factors which are subjective, and, according to Heisenberg, therefore, not subject to criteria of "true or false".

Planck, says Heisenberg, had put himself at the side of the Christian tradition, adding: "I doubt if human societies can live with such a net distinction between (scientific) knowledge and faith..." Wolfgang Pauli, shared his worries: "Inevitably —said Pauli—, everything will end up in tears. In the wake of religion all knowledge of the community did fit within a spiritual framework based, to a large extent in religious ideas and religious values". That framework had to be within the reach of the most simple members of the community even if the parables and figures of speech were only transmitting a vague idea of the subjacent values and concepts. According to Pauli, if the ordinary man is to live according to these values, he must be convinced that this spiritual framework embraces the full wisdom of this society. "To believe" does not mean "to take it for granted", but to have confidence in those values as a reliable guide. A society therefore finds itself in great danger when new knowledge menaces with exploding the old spiritual order.

The total separation of (scientific) knowledge and faith could be, at most, thought Heisenberg, an emergency measure, a temporary breathing moment. It is possible, said Pauli that in a non distant future, the parables and images of the religion may have lost their characteristic force even for the ordinary man. If this occurs, he said, "I am afraid that the old ethics will collapse as a castle of cards and then unimaginable horrors may take place". In summary, he said, I respect but I cannot subscribe Planck's philosophy. Similar considerations were devoted to Einstein's view points on the same subject.

Independently of the accuracy of Heisenberg's reminiscences on what Pauli said at that occasion about Planck's opinions on science and religion, it is very relevant to note that he was on target about the coming collapse of the old ethics and its progressive substitution for a new ethics almost indistinguishable from the old immorality.

As we enter into the third millennium of the Christian Era, two things seem to become increasingly evident: the widespread technological advances (based upon previous purely scientific advances) and a generalized retreat in religious practice, at least in the most opulent societies of Europe and America. Many have concluded a clear logical connection as cause and effect between both, assuming with no good reason that the increasingly rationalistic and godless mentality pervading modern culture has played an important role in the development of modern science.

Historically, this is not true. As shown convincingly by S.L. Jaki[2] the crucial first steps in the development of modern science took place in medieval European Christendom, and in no other place, when Buridan (1300-1358) and Oresme (1325-1382) introduced the idea of inertial movement[3]. Then Copernicus (1473-1543), Galileo (1564-1642), Kepler (1571-1630), and finally Newton (1642-1727) were able to pave the way to formulate a physics of motion capable of describing satisfactorily not only motions here on the Earth but the motions of the planets of our Solar System. Classical Mechanics set the rules for Optics, Electromagnetism, Thermodynamics, Chemistry and Modern Physics. In the wake of 20th century Planck and Einstein opened the way for Quantum Physics and Relativistic Physics which, within half a century, would produce, in quick succession, nuclear reactors, computers, lasers and space travel to the Moon.

Of course modern science, which has been able of describing satisfactorily the constituents of elementary particles as well as the behavior of the furthest galaxies located at billions of light years from our Milky Way, a science whose cradle was unmistakably in medieval European Christendom, acquired soon its proper autonomy and became independent of that Christian natural philosophy within whose premises was born.

Will modern science and technology be able to survive for very long after its metaphysical, epistemological, common sense foundations enter moral decomposition?

It is not at all clear.

Positive usefulness of Heisenberg's principle

Heisenberg's principle sets very strict limits to the accuracy with which pairs of conjugated variables (position and momentum, energy and time) can be known in a finite system. At the same time it makes very useful specific connections between them and therefore can be taken as a source of positive information about that system.

If to investigate photons or electrons we have at our disposal only such elementary particles as electrons and positrons, photons, baryons and antibaryons, it is no wonder that our potential for accurate knowledge is limited. And this applies one way or another to any finite system. But the universe itself is finite and in its early phase of expansion it was much more dense and much smaller that any ordinary elementary particle.

For instance, as noted in Chapter 7th and 8th, if one assumes that in its early phase of expansion the order of magnitude of the energy of the expanding universe was of the same order of magnitude as the uncertainty with which we know it, and that the order of magnitude of the time was the same as the order of magnitude of its uncertainty, we can write

$$\Delta E \cdot \Delta t \cong Mc^2 \cdot t \leq \hbar \tag{13.1}$$

Therefore

$$t = t_H = \hbar / Mc^2 \tag{13.2}$$

where M is the total finite mass (matter plus energy) of the universe, $\hbar$ is Planck's constant divided by 2π and c is the velocity of light in the vacuum.

The Schwarzschild radius R_+ of such expanding universe, which we may assume to be spherical would be given by

$$\frac{1}{2} mc^2 = \frac{GMm}{R_+} \tag{13.3}$$

As we have seen (Chapter 7th) if cosmic dynamics is described by Einstein's cosmological equations for an open universe

$$\dot{R}_+ = |k|^{1/2} c / \tanh y_+ \cong 2c \tag{13.4}$$

for $|k|^{1/2} \cong 2 \tanh y_+$, which corresponds to a cosmic density parameter

$$\Omega_+ = \frac{\rho(y_+)}{\rho_c(y_+)} = \frac{M\bigg/\dfrac{4\pi}{3}R_+^3}{3\big(\dot{R}_+ / R_+\big)^2 \big/ 8\pi G} = \frac{1}{2} \tag{13.5}$$

For a *flat* (Lambda Cold Dark Matter universe) M, of the order of 10^{12} galaxies made up by 10^{12} stars like our Sun, comes out relatively close to M for an *open* universe.

Heisenberg's principle is not against causality

Let us reproduce verbatim Jaki's word in "God and the cosmologists"[4]:

*Had Heisenberg's sensed something about the inherent limitations of the methods of physics, he would have refrained from stating that the inability of the physicist to measure nature exactly showed the inability of nature to act exactly. He would have needed only a modicum of philosophical sensitivity to note that the apparent truth of the foregoing statement depended on taking the same world **exactly** in two different meanings: one operational, the other ontological.*

*His confusion about those two different meanings meant an indulging in that elementary fallacy which the Greeks of old called **metabasis eis allogenos**. That fallacy was duly pointed out in the courses of introductory logic that were part of the curriculum in philosophy, without which no doctor's degree in any subject could be granted in German universities when Heisenberg received his own in physics in 1925.*

In taking Heisenberg's principle for a final disproof of causality, no qualms could be felt by physicists in the Anglo-Saxon world, where empiricism and pragmatism had for some time discredited questions about ontology... While Planck quickly perceived that the Machist interpretation of science threatens confidence in the reality of a causally interconnected universe, he invariably equated ontological causality with the possibility of perfectly accurate measurements.

Again, whereas in his three-decades long dispute with Max Born, Einstein often came to the defense of reality as such, he never perceived what he really wanted to defend. Nor is it likely that he would have been told about it as bluntly as by W. Pauli as the latter put the matter to Born in a letter written on March 31, 1954, in a room or two away from Einstein's own in the Institute of Advanced Study in Princeton.

Einstein's argument on behalf of a reality existing even when not observed amounted to no more than graphic phrases and gestures. He thought to the end that what he meant to be ontological causality, though he never called it such, could only be saved if perfectly accurate measurements were at least in principle possible. The chief failure of his famous thought experiments with a clock on a spring scale was not that it did not work but that it granted a reduction of the ontologically exact to exact measurability. He could not, of course, expect from Bohr, who refuted that thought-experiment, to be reminded of that reduction, since it was a principle with Bohr to avoid any reference to ontology as such.

When in 1987 "Physics Today" reported the spectacular burst of light and radiation form "Supernova 1987" hitting our planet, nobody had observed the light or the neutrinos coming from it. But they had been traveling towards the Earth at fantastic speed for many millions of years.

Does this mean that because nobody had observed it yet the explosion had not taken place yet?

Let us paraphrase Pascal in his famous Wager ("Pensees", 1662) about the existence of **objective physical reality**: yes or not.

*You must wager. It is not optional. Which will you then choose?... You have two things to loose, the **true** and the good; and two things to stake, your **reason** and your**will**, your **knowledge** and your **happiness**; and your nature two things to shun, **error** and **misery**. Your reason in not more shocked in choosing one rather than the other.*

*This is one point settled. But your **happiness**? Let us weight the gain and the loss in wagering that objective reality really exists. There are only two possibilities.*

*If you **gain**, **you gain all**; if you **lose**, **you lose nothing** because nothing exists, anyhow.*

*Wager then without hesitation that objective reality, the foundation of all appearances, **exists**.*

In other words two plus two make four: no more, no less.

It does matter whether it is **eight** or is **eighty**.

REFERENCES

[1] Werner Heisenberg, "Physics and Beyond" (Harper and Row: New York, 1971)

[2] Stanley L. Jaki, "The road of science and the ways to God" (University of Chicago Press: Chicago, 1978)

[3] Pierre Duhem Texts", selected by Stanley L, Jaki, in "Scientist and Catholic: Pierre Duhem" (Christendom college: Front Royal, Virginia 22630, 1991)

[4] Stanley L, Jaki, "God and the cosmologists" pp. 124-26, 130-31 (Real View Books: P.O. Box 1793, Fraser, Michigan, 1998). First published in 1989.

APPENDIX: FROM SCIENTIFIC COSMOLOGY TO A CREATED UNIVERSE

Almost exactly a hundred years ago, in 1885, Berthelot, a leading French scientist, declared that owing to the progress of science the world became free of mysteries. Then came a rash of unexpected discoveries: cathode-rays, X-rays, electrons, radioactivity and quanta. The world suddenly looked so mysterious that even politicians took note. On January 21, 1910, Jean Jaures, a leader of the French socialist party, told the Chambre des Députés: "The admirable scientist who once wrote that the world is without a mystery seems to me to have uttered a naiveté as great as his genius." Twenty years later there appeared the most successful high-level popularization of science published so far, *The Mysterious Universe* by James Jeans. The universe seemed to the famed British astronomer so mysterious that he was even willing to see beyond it a God, a sort of super-mathematician. A few years later Etienne Gilson wondered aloud why scientists should take any satisfaction in the apparent mysteriousness of the universe. Most Christian readers of the book seemed satisfied. Belief in God, the greatest mystery, often appeared safer when God's handiwork, the universe, also looked mysterious. Unfortunately, true mysteries were often seen in mere gaps of scientific knowledge, gaps which as a rule are rapidly filled as science progresses.

Even if by mysteries Berthelot meant such gaps, his really scientific bungle was not that he saw all mysteries, that is, gaps in scientific knowledge, being filled at least potentially, although this was enough of a bungle. In 1885 there was not even an inkling of the true explanation of spectral lines which by then had been measured by the tens of thousands and held the key to an atomic realm defying all scientific prognostication. Berthelot misread science most as he spoke of the universe without hinting at its scientifically problematic character. He should have known that in 1885 it was not yet possible to speak in a scientifically unobjectionable way about the universe, the totality of consistently interacting things.

Scientific cosmology was still to be born in spite of the fact that throughout the 19th century scientists spoke more and more often of the universe, a topic which had largely been the preserve of philosophers and of a few philosophically inclined scientists. That during the 19th century and before, scientists spoke of the universe in the sense of doing cosmology, was not in itself completely unscientific. One ought to speak of a problem one cannot solve if that problem is ever to be solved. The universe was for 19th century science, or for Newtonian science, a huge problem, but, and this was scientifically reprehensible, scientists preferred not to speak of this, rather, they tried to talk away the problem.

The problem was the alleged Euclidean infinity of the universe. Such a universe was largely the product of 19th century thinking. Newton himself believed the material universe to be finite in an infinite space. This idea was endorsed and propagated, as something most conformable to reason and to God, by no less publicists than Addison and Voltaire. But with the early 19th century there came a radical shift. Thus the astronomer Olbers declared in 1823, with a reference to Kant's authority, that the universe of stars had to be infinite. The context spoke for itself. Olbers wanted to save the infinity of the universe from the paradox of the darkness of the night sky. If the number of stars was infinite and homogeneously distributed, it followed, if one did not take into account the average lifespan of stars, that the intensity of starlight should be equal at any point to the brightness of the surface of a typical star, such as the Sun.

Whatever the loopholes in the optical paradox, which had already been discussed by Halley and intimated by Kepler, there were no loopholes in its gravitational counterpart. Here too the story goes back well before the 19th century. As early as 1692 Bentley called attention to the gravitational paradox in his famous Boyle lectures and also privately in his letters to Newton. After Green formulated in 1827 the theory of potential, the gravitational paradox could have been given a rigorous formulation. This did not happen until 1871 when Zöllner, professor of astrophysics in Leipzig, showed that in an infinite homogeneous universe any force obeying the inverse square law should produce an infinitely large potential at any point. At the same time Zöllner also suggested that Riemann's four-dimensional space-time manifold could provide a way out, provided the total mass of the universe was taken to be finite. Two years later, W.K. Clifford, professor of applied mathematics at University College in London, stated that Riemann's geometry made the universe a scientifically valid notion, the very basis of the possibility of a scientific cosmology.

Although both Zöllner and Clifford were prominent figures, scientists paid little if any attention. Some, like Seeliger in Munich, tried to change slightly the inverse square law to save the infinite homogeneous universe; others took the Milky Way for all the universe that was investigable. The infinitely large part beyond the Milky Way was declared by Kelvin, among others, to be forever beyond the reach of science. Such a solution was not science, but a schizophrenic thinking which split the universe into two parts: One finite (the Milky Way), the other infinite (everything beyond). Thus was the universe rid of its mysteries. So much in a way of commentary on the real scientific blunder which Berthelot made in speaking of the universe. Assuming as he did that the universe was infinite in the Euclidean sense, he had no right to speak of it as if it had no mysteries, that is, scientifically debilitating problems.

The first chapter in a scientifically rigorous cosmology came only in 1917 with Einstein's memoir on the cosmological consequences of General Relativity. There Einstein showed that the gravitational interaction of all material bodies could be given a formulation free from the gravitational paradox which plagued the so-called Newtonian universe. This was not, however, the most important aspect of that memoir which was the last in a series published by Einstein between 1915 and 1917 on General Relativity. The most important point was a formula which stated that from the values of the average density of matter and of the gravitational constant one could infer the value of the total mass of the universe and its overall radius of curvature produced by that mass. Such an inference was not a mere play with formulas. The inference was based on a broad theory, General Relativity, which already at that time provided three experimentally verifiable predictions, each independent of the other: the gravitational red-shift, the gravitational bending of light, and the advance of the perihelion of planetary orbits (readily observable in the case of Mercury). While the early verifications of these effects were not altogether convincing, today the margin of error in measuring them is too small to permit real doubt. During the last half a century several other predictions of General Relativity have been submitted to observational tests which further confirmed its reliability. No less valuable confirmation of General Relativity is the increasingly vaster field of study: scientific cosmology. All branches and further developments of that study are based on the conviction that it is scientifically meaningful to discuss the consistent interaction of the totality of matter, a totality which is the universe.

Such a conviction is further strengthened by the fact that in modern scientific cosmology the study of galaxies and other large-scale celestial objects is closely united with the atomic, nuclear, and subnuclear studies, that is, the study of the smallest material objects. That scientific cosmology deals with the totality of consistently interacting matter is strikingly shown by the expansion of the universe. This large-scale motion to which all matter is subject was first a mere prediction by Abbe Lemaitre on the basis of General Relativity before it obtained observational proof in the red-shift of the spectrum of galaxies. That shift can be explained only if galaxies have a recessional velocity with respect to one another which is the greater the farther they are from one another. This, however, entitles one to follow the motion of galaxies backward in time to a moment where all matter was condensed within a relatively small space.

Scientific cosmology gives so far an account only of the gravitational interaction of matter. Other interactions, electromagnetic and nuclear, are incorporated only in part in that picture. A Unified Field Theory valid for all these forces (to say nothing of some forces still to be discovered) is still to be formulated. Such a theory is the most coveted prize for leading scientists today, in proof that in their eyes the notion of the universe is a truly valid and not merely a regulative idea. They are encouraged by the success achieved so far by General Relativity. Twentieth-century science provides indeed powerful support on behalf of the validity of the notion of the universe, a validity which since Kant's *Critique of Pure Reason* has been largely dismissed in philosophical circles (and even in some theological realms which should know better). Kant's ultimate purpose with the *Critique* was, it is well to recall, to provide a rigorous basis for man's autonomy or independence. In that strategy an all-important step was the presentation of the universe as a bastard product of the metaphysical cravings of the intellect. If the universe was not a valid notion, then it could no longer function as a reliable basis of inference to the existence of the Creator, the only Independent

Being. In the absence of such trustworthy inference man's autonomy seemed to be fully secured.

The really dangerous part in Kant's strategy was that he largely succeeded in creating the impression that his reasoning was in the spirit of exact, that is, Newtonian science. Now, if he had been well informed and really perceptive in matters scientific, he should have realized that his antinomy about the finiteness versus infinity of the universe had no scientific merit. The so-called Newtonian infinity, which he had in mind, had no scientific validity and therefore could not be used as an alternative to the possible finiteness of the universe. Twentieth-century science or cosmology has shown that it is meaningful to speak about the infinity of the universe as well as about its finiteness, without being thrown back thereby into the hold of Kant's first antinomy. That modern scientific cosmology restores confidence in the validity of the notion of the universe should seem no small bonus for those who see the universe as a jumping board to its Creator. The same cosmology does something even more important in that respect. The universe as described in that cosmology strikes us as a truly existing thing. Scientific cosmology provides this impression not by philosophical arguments, however valid and precious they may be about existence. Rather, it does so by its portrayal of the universe as a most specific, most peculiar, most particular and at the same time fully consistent entity.

Here again a recall of the historical background against which this development should be seen may be very helpful. A hundred years ago Herbert Spencer rode the crest of the wave with his *First Principles*, a cosmic philosophy. Like Kant, Spencer too succeeded in presenting his philosophy as steeped in science. Spencer first made a name for himself with his account of Laplace's nebular hypothesis in which the solar system, and ultimately all solar systems and celestial bodies, were the products of an evolution which started with a nebulous, that is, most homogeneous state of matter and ended with a most specific or inhomogeneous form of it. The effort was a somersault both philosophically and scientifically, worthy of a philosopher who did not have proper scientific training and who prided himself that never in his life was he surprised by anything. Yet, because Spencer's scientific bungle was not perceived, his philosophical bungle, too, was readily overlooked. The scientific bungle was tied to the lack of any information about the nebulous state, the presumed starting point of cosmic evolution. The pale whiteness of nebulae was simply taken by scientists for a homogeneous fluid for no real reason whatsoever. Scientific unanimity based on wishful inference fully reassured Spencer that his starting point was correct and reliable. For all that, it remained thoroughly incorrect for him to assume that complete homogeneity would ultimately give rise to a high degree of inhomogeneity. As one would expect, Spencer assumed a "very slight imbalance" in that original state, but the significance of this proviso was largely lost on him and on his readers, among them Charles Darwin who naively took Spencer for one of the greatest intellects of all times. Spencer, as is well known, took that homogeneous, nondescript, unspecified entity for the starting point of the natural form of existence. It was a matching counterpart to the perceptual non-specificity of Euclidean infinity which in the eyes of many served as a natural frame of existence which needed no further, let alone metaphysical, explanation.

A hundred years after Spencer the farthest point in the past to which scientific cosmology carries us is the very opposite of non-specificity. With the discovery in 1965 of the 2.7° K cosmic background radiation, a proof was served on behalf of earlier theories about the genesis of chemical elements, at a time when the universe was in a highly condensed state. Chemical elements, as ranged in the Mendeleev table, are not only very specific in themselves (Mendeleev himself was so impressed by their specificity as to take all of them for irreducible primordial entities). Their relative abundances are no less specific. Such a compound specificity could arise from a cosmic soup comprising all matter in which for each proton, neutron, and electron there were almost exactly 40 million photons at a very specific temperature and pressure. Only under such conditions could the interaction between those particles yield hydrogen and helium in their actual and very specific proportion and make thereby the genesis of heavier elements possible.

On looking at a proportion of 1 versus 40 million, no sane mind would be tempted to take it as a natural or, rather, exclusive state of affairs. The same lesson is on hand when one follows modern scientific cosmology beyond the baryon state of the cosmos to states which modern scientific cosmology calls lepton, hadron, and quantum states. Beyond those states is the matter-antimatter state where things appear dizzyingly specific. Nothing would be more natural than to see that state as comprising an equal amount of matter and antimatter. But interaction between equal amounts of matter and antimatter would yield only sheer radiation. In order that our ordinary matter and world may arise, scientific cosmology must resort to a most "unnatural" assumption, according to which there had to be an original imbalance of one part in 10 billion in favour of ordinary matter. Such is not a mere speculation. Its reasonability is implied in the finding by Fitch and Cronin of a slight asymmetry in the decay of K_2 mesons, a finding which earned them the Nobel Prize in 1980.

Clearly, there is an immense contrast between the primordial state of matter as described by Spencer and as described by scientific cosmology, a contrast which should provoke the utmost surprise. The difference is not merely a difference between studies vagueness and a study steeped in scientific precision. The real difference is that a most specific entity may strike even a philosophically desensitized mind with the fact of existence. While the queer specificity of everyday things can easily be lost on us, such is hardly the case when one is forced to face up to cosmic specificity, described and verified in all details by science. About the universe scientific cosmology states not only that it is a valid notion, but also powerfully suggests that it does exist — a most welcome contribution in an age in which philosophizing is stranded on the shallows of idealism and logicism, two skillful guides to solipsism and sheer willfulness.

To grasp all this requires no familiarity with that elaborate mathematics which is an integral part of scientific cosmology. Because of its mathematical aspect, scientific cosmology is subject to Godel's incompleteness theorem, according to which no set of non-trivial arithmetical propositions can have its proof of consistency within itself. This means that all scientific efforts aimed at an account for the universe, which would show that the universe can only be what it is and cannot be anything else, are doomed to failure. Eddington was not the only major scientist in our century who seriously devoted himself to such an undertaking. Einstein himself would have loved to construct a Unified Theory such that, as he put it half seriously, "even the good Lord would not be able to improve on it." In the past two decades several Nobel laureates admitted that their work was motivated by some such aim. Of course, it is not absolutely beyond the realm of possibility that a scientist should be fortunate enough to hit upon a mathematical formalism which would fit the quantitative aspect of all material processes. In that case there would remain no mysteries or unsolved problems with respect to the physical universe where, let us not forget, God "disposed everything according to measure, number, and weight." Such a fortunate theory would account not only for all data on hand but also for data still to be gathered in the future, however distant.

Yet even such a theory could not claim to itself intrinsic consistency. Its proof of consistency would, in virtue of Godel's theorem, lie in a set of considerations not included in it. In other words, scientific cosmology, because of Godel's theorem, can never pose a threat to that cosmic contingency which is intimated in the scientific portrayal of the specificity of the universe. A universe which is contingent is the very opposite of cosmic necessitarianism, the age-old refuge of materialists, pantheists and atheists, all of whom, with Nietzsche in the lead, consider the dogma of creation as the most pernicious error man can espouse.

The final and most striking pointer of scientific cosmology to the createdness of the universe is a sequel to the contingency of the universe. The contingency meant here is not its confused sense equivalent to an undefined indeterminacy. Contingency here means the utter dependence of something on something else. The actual specificity of the universe is a striking reminder of such a dependence. Precisely because the actual cosmos is so specific, it should be easy to see the possibility of an immensely large number of other specificities. The actual specificity of the universe, which cannot be necessary, reveals therefore its dependence on a choice beyond the universe. Since the specificity of the universe is highly understandable, the choice underlying that specificity, a choice which also gives the universe its actual existence, must involve an intelligence and power which is supercosmic, that is, beyond that cosmos which for science is the totality of consistently interacting things. Things, even world, which do not interact consistently are, it is well to recall, irrelevant for science. Nor is relevant for science that spurious philosophy which is often equated with quantum mechanics, the probabilistic method to account for atomic and subatomic phenomena. The radical inconsistency or purely chance character which is attributed by that philosophy (Copenhagen theory) to atomic processes, is a consequence of the radical rejection by that philosophy of any question about being (ontology). Typically, such a philosophy is not consistent to the point of recognizing the fact that it therefore has no right to ask, let alone to answer, the question: What is chance?

Is it reasonable to assume that an Intelligence which produced a universe, a totality of consistently interacting things, is not consistent to the point of acting for a purpose? To speak of purpose may seem, since Darwin, the most reprehensible procedure before the tribunal of science. Bafflingly enough, it is science in its most advanced and comprehensive form — scientific cosmology —

which reinstates today references to purpose into scientific discourse. Shortly after the discovery of the 2.7°K radiation cosmologists began to wonder at the extremely narrow margin allowed for cosmic evolution. The universe began to appear to them more and more as if placed on an extremely narrow track, a track laid down so that ultimately man may appear on the scene. For if that cosmic soup had been slightly different, not only the chemical elements, of which all organic bodies are made, would have failed to be formed. Inert matter would have also been subject to an interaction different from the one required for the coagulation of large lumps of matter, such as protostars and proto-solar systems.

Yet the solar system ultimately emerged and with it that curious planet, the Earth, which if placed at a slightly different distance from the Sun, would have undergone a very different evolutionary process on its surface. At any rate, the emergence of life on earth is, from the purely scientific viewpoint, an outcome of immense improbability. No wonder that in view of this quite a few cosmologists, who are unwilling to sacrifice forever at the altar of blind chance, began to speak of the anthropic principle. Recognition of that principle was prompted by the nagging suspicion that the universe may have after all been specifically tailored for the sake of man.

That scientific cosmologists were forced by their own findings to formulate the anthropic principle may please some philosophers and theologians. In Aquinas' philosophy it was a central tenet that the universe was created for the sake of man. It must not however be forgotten that such a tenet, or the anthropic principle, can never be a part of scientific cosmology. Science is about quantitative correlations, not about purpose. Not that science as such is not a purposeful activity. As all truly human actions, science too, is for a purpose and to a very high degree. This is true even of those scientists who devote their whole lives to the purpose of proving that there is no purpose. Such scientists, as Whitehead once put it, constitute an interesting subject for study. And yet, no matter how deeply is the actual implementation of scientific method steeped in purpose and therefore steeped in metaphysics, it is very important to keep in mind the self-imposed limitations of that method. Otherwise one will expect from that method something it cannot deliver. Scientific cosmology can reassure the philosopher that science poses no threat to the validity of such notions as universe, existence, and contingency. Actually, scientific cosmology powerfully suggests these notions and indeed makes use of them on a vast scale. But a suggestion, however powerful, is one thing, philosophical demonstration is another. While science or scientific cosmology can be a powerful prompting for considering the createdness of the cosmos, it can never become a discourse about creation as such.

The importance of this distinction becomes obvious when creation in time is considered. If there is a theological theme, it is creation in time, the theme or dogma which supports all other Christian themes and dogmas. Whenever the meaning of creation in time is weakened, let alone eliminated, the meaning of all other tenets of the Christian creed become weakened or eliminated. Those tenets — Fall, Incarnation, redemption, the growth of the Kingdom of God, eschatology, final judgment— presuppose not only creation but also a creation in time because all those tenets refer to events in time which alone can constitute that sequence which is salvation history When in 1215 the Fourth Lateran Council solemnly defined creation out of nothing and in time as a dogma, it merely confirmed a long-standing tradition. The continued strength of that tradition, which, by the way, was again reasserted by Vatican I, is so great as to evidence itself far beyond Christian realms. A case in point is the widespread custom of scientists and science writers to refer to the dating by science of the age of the universe. Few customs can become more unscientific. While science can assure us that it can carry its investigations 12 billion years back into cosmic past, there is no science whatever which can date the birth of the universe. There is no scientific value whatsoever in statements, often seen nowadays in print, that through the launching in 1985 of the Space Telescope man will have a glimpse of the moment of creation, because his farthest view into the universe will be increased from 2 to 20 billion light years. The reason for the absence of science in such statements is simple. Physical science or scientific cosmology is absolutely powerless to show that any stage of material interactions is not reducible to a previous state, however hypothetical. If science is impotent in this purely scientific respect, it is even more impotent with respect to a far deeper problem, a problem of very different nature, namely, that a given physical state must owe its existence to a direct creative act, which brought that physical state into being out of nothing.

Scientific cosmology has, however, made a very important contribution with respect to the existence of time, the very basis for making meaningful the phrase, creation in time. Scientific cosmology shows all too well that the universe carries on itself the stamp of time. Such a stamp is the expansion of the universe. In a very real sense the universe is ageing. It clearly burns up energy and by doing so it shows the signs of transitoriness. The force of that sign can best be judged by the frantic efforts of some cosmologists to erase that stamp from the face of the universe. The enthusiasm with which the steady-state theory was hailed thirty years ago is a case in point. The real aim of that theory was to secure for the universe that infinity along the parameter of time which it already lost along the parameter of mass and space. That the theory was indeed markedly antitheological in character could easily be gathered from the emphatic insistence of its proponents. They claimed that the continued emergence of hydrogen atoms, whereby the density of matter is kept steady in an expanding space, should be conceived as a creation out of nothing though without a creator. For the atheistic candor of those proponents one ought to be appreciative. It is hardly to be expected that they would be appreciative of the remark already made about the impossibility of physics to see the nothing beyond any given state of matter.

An equally atheistic, or simply pagan, or at best agnostic, longing for the eternity of matter is beneath that jubilation which greets the periodic news about the finding of the so-called missing mass, a curiously countertheological counterpart of the still elusive missing link. If that extra mass should be found in cosmic spaces, then the present expansion would turn into a contraction, and possibly that contraction would be followed by another expansion. Yet, even in this case the process would not go on ad infinitum. There has not yet been found any physical process that would be exempt from the law of entropy. Indeed, more and more attention has been given recently to the rate at which subsequent cycles in an oscillating universe would be less and less energetic. It is indeed possible to calculate, however tentatively, the number of cycles which would bring us back in time to the point where the period of a cycle would be vanishingly small. To some sanguine souls and uninformed minds that vanishing point may appear the moment of creation which would then certainly vanish.

The idea of an oscillating universe presupposes the finiteness of matter. That finiteness, when cognizance was first taken of it in the early 1920s under the impact of Einstein's General Relativity, produced shock waves in scientific and philosophical circles in which the infinity of the universe had for some time played the role of a convenient ultimate entity, making God unnecessary. The shock waves were all the more telling because, as Einstein already pointed out in 1917, there were ways in which it is possible to assume the infinity of matter without running into contradictions. Yet, all those ways are such as to provide further evidence on behalf of the stunning specificity of the universe. A distribution of infinite matter which would give rise to paths of motion resembling a cylindrical helix is too specific to be taken for that natural and necessary form of existence for which Euclidean infinity could so readily pass. The same is true about a distribution of infinite matter which would permit motion only along the curving slopes of a saddle with no edges, corresponding to an infinite hyperbolic space. About none of these specific situations is it natural to say that they are such forms of existence which one would naturally expect to exist and exist necessarily at the necessary exclusion of all other possibilities.

So much, in broad strokes, about the contribution of scientific cosmology to the idea of a created universe. The suggestiveness of that contribution is anything but small. Long before the discovery of the 2.7° K cosmic background radiation filled the world of science with metaphysical puzzlement, there was plenty of it under the surface. Einstein indeed felt it necessary to reassure with the words, "I have not yet fallen in the hands of priests," a friend who worried that on account of his cosmology Einstein might become a believer.

Reluctance to face up to the fact that the universe has a message pointing beyond itself is an old story. All too often the reluctance issues in a patently antiscientific posture. John Stuart Mill, who saw in cosmology the stronghold of theists, did his utmost to discredit it. In the process he deprived the cosmos of its intrinsic rationality. He did so by peddling the idea that in some faraway regions two and two may not necessarily make four.

Since Mill the same story has been enriched with further and no less telling chapter. They are usually provided by those educated in a milieu in which "interest in the greatest problems that ever agitated man is successfully stifled." Such was Henry Adams' characterization of that intellectually high-powered milieux of Boston and Harvard where he was brought up. Bologna, Paris, Oxford, Cambridge, Bottingen, Uppsala, Basel, Leiden — to keep the historical order — and many other illustrious places of learning, would provide ample material for painting that milieu. What is stifled is not, however, extinguished. Henry Adams had to realize, fifty years after he left Harvard, that "if he were obliged to insist on a Universe, he seemed driven to the Church." So he opted for what he called the Multiverse. He did not suspect that his option for the counter-metaphysics of multiworlds demanded a renouncing of science at its best. The coming of scientific cosmology was less than a decade away from the moment when Henry Adams looked for salvation in multiworlds which, precisely because they could not interact consistently, could not form a universe and were therefore useless for science.

Science and Universe form indeed a seamless garment, a thesis not falsifiable unless the principle of falsifiability is turned into a *petitio principii*. That all science is cosmology has been an old truth long before K.R. Popper, hardly suspect of metaphysics, earned the aura of originality by voicing that truth to an unsuspecting generation which failed to notice its exemptness, implied by him, to the unrestricted sweep of falsifiability. Science, philosophically and historically, is an ally, not of the Academy of agnostics but of that Church which, unlike some of her theologians, knows all too well why her creed starts with the words: "I believe in God, the Father Almighty, Maker of Heaven and Earth." The effort which tries to resolve conflicts between Christianity and science by stating that religion is about persons and not about the universe of things, should seem a very poor half truth. For God, at least the Christian God, is above all the Creator of the Universe. Thanks to science, that universe appears less and less mysterious, though at the same time more and more specific, and thereby an irrefragable pointer to God, the mysterious origin of all.

Notes

For further details and documentation of the main topics of this paper, see my latest books, *Cosmos and Creator* (Edinburgh: Scottish Academic Press, 1981; Chicago: Regnery Gateway, 1982) and *Angels, Apes and Men* (La Salle, ILL.: Sherwood Sugden and Company, 1983). The history of the optical and gravitational paradoxes is given in my book, *The Paradox of Olbers' Paradox*

(New York: Herder and Herder, 1969) with additional data in my article, "Das Gravitations-Paradoxon des un- endlichen Universums," *Sudhoff's Archiv*, 63 (1979), pp. 105-22. The historical context of Godel's theorem and its first application to physics and cosmology can be found in my book, *The Relevance of Physics* (Chicago: University of Chicago Press, 1966), pp. 127-31. On the half-a- century-old history of the antiontological meaning attributed to chance in the Copenhagen school, see my article, "Chance or Reality: Interaction in Nature versus Measurement in Physics," *Philosophia* (Athens), 10-11 (1980-81), pp. 85-105. The utterances of H. Adams are from *The Education of Henry Adams* (New York: The Modern Library, 1931), pp. 34 and 429.

CHRONOLOGY

Werner Heisenberg (1901-1976)

1901 Werner Karl Heisenberg is born in Wurzburg (Germany) the 5 of December.

1920 Enters the University of Munich and the Seminar of Arnold Sommerfeld.

1923 Ph. D. at the University of Munich, Becomes assistant of Max Born at the University of Gottingen.

1925 Heisenberg, Born and Jordan publish a seminal paper on the ground state and excited states of electrons in atoms and the quantum jumps between them through the absorption or emission of light.

1927 Heisenberg publishes his famous uncertainty principle.

1928 He is appointed professor of Theoretical Physics at the University of Leipzig.

1932 Sets forth his quantum model of the atomic nucleus in which neutron and proton can be viewed as two different quantum states of the same elementary particle.

1933 Receives the Nobel Prize in Physics 1932, for his contribution to set forth the formal basis of Quantum Mechanics.

1937 Contracts matrimony in Berlin with Elisabeth Schummacher.

1939 At the beginning of World War II becomes involved in the German nuclear project.

1942 Is named director of the Institute of Physics Kaiser Wilhelm in Berlin.

1943 Is appointed professor of Physics at the University of Berlin.

1945 At the end of World War II the Allies bring him to Farm Hall (England) in July 1945.

1946 He becomes director of the Max Planck Institute of Physics and Astrophysics I Gottingen.

1951 He is named president of the German Atomic Energy Commission and of the German delegation for the constitution of the CERN.

1976 Dies of cancer at his home in Munich the 1st of February.

Georges Lemaitre (1894-1966)

1894 Georges Lemaitre is born in (Belgium) the 17 of July.

1914 Begins studying Civil Engineering at the Catholic University of Louvain. Interrupts his studies to serve as Artillery officer in World War I.

1918 Begins preparation for priesthood.

1920 Obtains his doctorate with a Thesis entitled "L'Aproximation des functions de plusiers variables reales... "under Professor Ch. De la Vallee Poussin.

1923 Is ordained catholic priest. Spends a year at the University of Cambridge, UK working with Arthur Eddington on Cosmology, Stellar Astronomy and Numerical Analysis.

1924 Spends a year at Harvard College Observatory, Cambridge, Massachusetts with Harlow Shapley, and at MIT.

1925 Becomes part time lecturer at the Catholic University of Louvain, Belgium.

1927 Publishes his work "Un universe homogene de masse constant et de rayon croissant rendant compte de la vitesse radiale des nebulenses extragalactiques" where he derived Hubble's Law.

1931 Lemaitre translates into English his article with the help of Arthur Eddington but the part estimating the value of Hubble's "constant" is not translated. At this time Einstein tells Lemaitre "your calculations are correct but your physics is abominable".

He obtains his Ph. D. at MIT and becomes ordinary professor at the Catholic University of Louvain.He is invited to lecture in London at the meeting of the British Association and publishes his report on the "Primeval Atom in Nature. Later Fred Hoyle strongly critizices Lemaitre's theory calling it the "Big Bang Theory" and defends the "Steady State Theory", which postulates continuous creation of matter, supported by himself, Bondim, Gold.

1933 Lemaitre and Einstein travel together to California to give a series of seminars there.

1934 Lemaitre receives the highest Belgian decoration form King Leopold. A. Einstein, Ch. De la Vallee Poussin and A. Hemline are his scientific sponsors.

1936 He is elected member of the Pontificia Academy of Sciences

1951 He disagrees privately with a famous discourse of Pope Pious XII before the Pontificia Academy of Sciences on the Proofs of God's existence in the light of Natural Modern Science.

1960 He is named president of the Academy during the Pontificate of Pope John XXIII

1964 He suffers a heavy heart attack.

1966 He dies the 20th of June shortly after having news of the discovery of Penzias and Wilson reporting the discovery of the Cosmic Background Radiation clearly compatible with the Big Band theory and very difficult to reconcile with the Steady State theory.

GLOSSARY

Alpha particle. Nucleus of ^{4}He atom made up of two neutrons and two protons. After primeval nucleosynthesis the cosmic plasma is made up mainly by protons (76%), α particles (24%) and electrons.

Baryogenesis. Process by which a definite population of baryons (protons, neutrons) becomes well defined in the early universe. This requires that the cosmic matter density becomes equal or less than nuclear density, which occurs at a cosmic temperature $T_b \approx 3.88 \times 10^{12}$ K. It must have taken place at $t \sim 10^{-5}$ s.

Baryon-to-photon ratio. At present the universe is *transparent* and the ratio of baryons to photons in the universe is $n_b / n_r \approx 10^{-9}$ and is kept constant. It is to be expected that before atom formation, at temperatures above 3000 to 4000 K, the universe was in an *opaque* plasma state, in which there was considerable friction (multiple scattering) and the density of photons increased with time with respect to the density of baryons (nuclei of H and ^{4}He).

Big Bang nucleosynthesis. Primordial process by which cosmic protons and neutrons fuse together to form ^{4}He nuclei and traces of other light nuclei. This occurs at a temperature $T_{ns}\sim$ 4.60×10^8 K. It must have taken place at $t \sim 10^3$ s.

Black hole. A physical system so massive and compact that in it the strong gravitational field prevents even light from escaping. The primordial cosmos could not have been a black hole because in that case the big bang could not have taken place.

Blackbody radiation. A blackbody is an object which is in thermal equilibrium at a given temperature and emits radiation according to Planck's formula. The intensity of the emitted radiation depends smoothly on frequency at any given fixed temperature and presents a characteristic maximum.

Blueshift. If a star or galaxy moves toward us the radiation emitted by it observed from the Earth appears shifted toward shorter wavelength (towards the blue) (See Doppler shift).

Cepheid variable. A type of variable star whose brightness was seen to vary over time. Henrietta Swann Leavitt discovered in 1912 that an approximately linear relationship existed between the period of a Cepheid's pulsation and its luminosity. The longer the period of its pulsation, the greater was the star's luminosity. Once the distance to nearby Cepheids was determined, they became useful measuring sticks for establishing the distances to other galaxies after they were discovered. In 1925 Edwin Hubble discovered a Cepheid in M31, the Andromeda Nebula, which enabled him to estimate that it was 8×10^5 light-years from Earth. It convinced astronomers that Andromeda was not part of the Milky Way, but a galaxy in its own right.

COBE. Cosmic Background Explorer, an orbiting satellite with three scientific instruments on board.

Cosmic accelerated expansion. Recent observations of magnitude (related to distance) versus redshift (related to recession velocity) which indicate that the rate of change of velocity with distance is larger for nearby galaxies than for far-away galaxies. To properly interpret the evolution in rate of change it is necessary to take into account that the maximum recession velocity cannot exceed the speed of light and that for the most distant protogalaxies relativistic effects must be expected.

Cosmic density parameter (Ω). Dimensionless parameter giving the ratio between actual mass density (ρ_m) and critical mass density (ρ_c), corresponding to cosmic matter moving barely at escape velocity. $\Omega(t) = \rho_m(t)/\rho_c(t)$ is time dependent, and, at present, using local data $\Omega(t_0) = \rho_m(t_0)/\rho_c(t_0) \approx 0.04$, very small in comparison with $\Omega(t_{af}) \approx 0.99$, corresponding to the time of atom formation ($t_{af} \approx 300000$ yrs).

Cosmic mass. Einstein's cosmic general relativistic equations are consistent with a very large but finite mass for the entire universe. Present cosmic dynamics allows one to estimate the total cosmic mass as $M_U \approx 4.5 \times 10^{54}$ g. Note that this number if of the order of $N_G N_S M_{Sun}$ where $N_G \sim 10^{11}$ (galaxies) times $N_S \sim 10^{11}$ (stars) times $M_{Sun} \sim 2 \times 10^{33}$ g, the mass of a typical star (the Sun).

Cosmic time parameter $H \times t$. Cosmic parameter (dimensionless) giving the product of the Hubble's parameter $H = \dot{R}/R$ and the cosmic time at any moment in the cosmic expansion. For an open universe $H \times t$ is less than one and more than two thirds (2/3). This parameter is time dependent, because H is time dependent and, of course, t is also time dependent. At present, using local data, $H \cdot t \approx 0.94 \pm 0.06$, substantially larger than $H_{af}t_{af} \approx 0.667 \approx 2/3$, corresponding to the time of atom formation ($t_{af} \approx 300.000$ yrs).

Cosmological constant. A term in Einstein's cosmological equations of general relativity which results in repulsive interaction with opposite sign to the gravitational attraction. Inflationary theories interpret this term as a measure of the energy density of the vacuum.

Dark energy. Since $\Omega \approx 1$ is consistent with the estimated matter-energy density at the time of "decoupling" ($T \approx 3000$ K), and the dark mass has been taken as contributing about 30% to it, the difference (about 70%) is commonly attributed to the repulsive force of the vacuum (cosmological constant) in the form of a kind of potential "dark energy": For a discussion of the time dependence of the matter mass density ($\Omega_m(t)$) and the potential space-time curvature energy density ($\Omega_k(t)$) resulting in $1 = \Omega_m(t) + \Omega_k(t)$, is compatible with $\langle \Omega_m \rangle \approx 0.26$, $\langle \Omega_k \rangle \approx 0.74$ averaged from the time of galaxy formation to the present.

Dark matter. In the close neighborhood of our galaxy the matter mass density in galaxies is only about 4% of the critical density (required for galaxies being exactly at *escape* velocities). It is commonly assumed that in order to have $\Omega = \rho/\rho_c = 1$ over cosmic space-time a matter mass density of at least 30% is required, including matter of non-baryonic nature. This is the so called "dark matter". In discussing dark matter, the time dependence of $\Omega(t)$ from very early times to present as given by the Friedmann-Lemaitre solutions of Einstein's cosmological equations, is usually ignored.

Dark night paradox. The paradox which points out the apparent impossibility of having a dark night in a universe with an infinite number of luminous stars homogeneously distributed.

Density perturbations. Fluctuations of the density of matter or radiation in the early universe which can be later amplified by gravity resulting in proto galaxies at early cosmic times.

DIRBE. Diffuse Infrared Background Experiment. One of the three COBE instruments, cooled at 1.5 °K by liquid helium and designed to measure any cosmic primordial infrared radiation as well as any other infrared radiation from stars and dust in the Milky Way and other galaxies. It works at 10 different wavelengths from 1.2 to 240 micrometers.

DMR. Differential Microwave Radiometer. One of the three COBE instruments designed to look for small anisotropies in the cosmic microwave background radiation; it works at 3 wavelengths, 3.3, 5.7 and 9.6 mm. The 3.3 and 5.7 mm receivers operate at 140 °K, and the 9.6 mm works at room temperature.

Doppler shift. Shift in the receiver frequency (and wavelength) of waves (sound or electromagnetic radiation) due to the relative motion of source and observer, depending on their approach or recession. For light, i.e., electromagnetic radiation, results in a blueshift or a redshift, respectively. Christian Doppler first proposed this effect in terms of the change in pitch for sound waves in 1841, but modern astronomers apply it to electromagnetic waves, corresponding to an observed change in color. Stars or galaxies with spectra shifted to the blue (or shorter) wavelengths of the spectrum are moving toward the Earth. Stars and nebulae with spectra shifted to the red, or longer, wavelengths are moving away.

Electromagnetic interactions. Interactions due to electric charges. Static charges give rise to electric fields, accelerated/decelerated charges produce electromagnetic waves, including radio, microwave, infrared, visible, ultraviolet light and X-rays, as well as γ-rays.

Electroweak interactions. Unified description of weak interactions (responsible for nuclear β decay) and electromagnetic interactions, due to Glashow, Weinberg and Abdus Salam (1967 -1970).

Energy conservation. Formulated by nineteenth century physicists (originally by Julius R. Mayer in 1842). It is the principle that establishes the strict *conservation* of energy: energy is not destroyed or created in physical processes, it is only transformed, for instance, thermal energy into mechanical energy, or vice versa. After the advent of Relativity Theory, associating an energy $E = mc^2$ to any mass m, the principle was generalized to include the energy associated to rest mass.

Expansion of the universe. The expansion of cosmic space suggested by Einstein's field equations of general relativity. Alexander Friedmann showed in 1922 that such an expansion was a natural consequence of the field equations, but Einstein, at first, disagreed with Friedmann, who died in 1925.

Finite universe. Einstein and Lemaitre, among others, assumed that the universe was finite because otherwise logical contradictions are unavoidable. This is the only satisfactory way of explaining Olbers' paradox. The last generation of powerful telescopes would be able to detect galaxies substantially further away than the most distant galaxies presently observable but they do not.

Finiteness of the universe. We assume the universe to be intelligible (otherwise we would not be investigating it). It is *infinite* or *finite*. If it is infinite it is beyond our capability of understanding. So we *conclude* it is finite, and hope not to incur in contradiction, as pointed out explicitly by Georges Lemaitre, and as required explicitly by Einstein.

FIRAS. Far Infrared Absolute Spectrometer. One of the three COBE instruments, designed to measure the spectrum of the cosmic background radiation and to compare it with the predicted blackbody Planck distribution curve; it is cooled at 1.5 °K by liquid helium and measures wavelengths from 0.1 to 10 mm by means of a very precise interferometer.

Flat universe. A homogeneous, isotropic universe just at the borderline between spatially *closed* (with a halt in the expansion followed by contraction) and spatially *open* (expanding for ever with non zero acceleration). The geometry for a flat universe is precisely Euclidean. The general relativity cosmological equations are consistent with a flat universe only if the total mass of the universe is infinity, which brings forth the spectre of Olbers gravitational paradox.

Flatness problem. A problem, real or rhetorical, of traditional big bang theory pointed out by proponents of cosmic inflation which requires that $k = 0\,(\Omega = 1)$ always, regardless of observational evidence to the contrary in our local neighborhood $(R \sim R_0,\ t \sim t_0)$.

Gauge theories. Theories which allow transformations in the dynamical equations leaving invariant the scalar and vector potentials describing physical interactions. The theoretical equations describing electromagnetic interactions (QED) are an example of gauge theory. The theoretical equations describing the strong nuclear forces, i.e. quantum chromodynamics (QCD) equations, are an example of gauge or Yang- Mills theory resulting in the so called asymptotic freedom for the way quarks are bound within the nucleons.

Gluons. Particles which play a role in the strong interactions analogous to that of photons in electromagnetic interactions.

Gödel's theorems. Fundamental theorems due to Kurt Gödel which state that in any non-trivial mathematical (or physical-mathematical system) the statement which is its proof of consistency must be outside the system. As a consequence there cannot be a "final" physical-mathematical theory.

Gravitation. The mutual attraction between any two massive particles. In a cosmic scale gravity appears as strong because it has an infinite range, like the electromagnetic interaction, and unlike the weak and strong nuclear forces, but is always attractive. Only in inflationary theory a cosmic false vacuum is postulated that results in a repulsive force resembling a negative gravitation.

Heisenberg's principle. Relationship between pairs of conjugated variables (p = momentum, x = spatial coordinate, E = energy, t = time) which set strict limits to the accuracy with which they can be known simultaneously:

$$\Delta p \cdot \Delta x \geq h \text{ and } \Delta E \cdot \Delta t \geq h$$

Heisenberg's principle can be positively useful to get information on physical interactions involving photons, neutrinos and elementary particles.

Horizon problem. A problem, real or rhetorical, of traditional big bang cosmology pointed out by proponents of cosmic inflation. If at very early times after the big bang the universe was homogeneous and isotropic to start with, the horizon problem disappears altogether.

Hubble parameter. The ratio of recession velocity of galaxies ($\dot{R}$) to distance (R), improperly called Hubble's constant, because for nearby galaxies $H_0 \approx \dot{R}/R$ is approximately constant, but, in principle, it is time-dependent, being $H(t) \approx \dot{R}(t)/R(t) >> H(t_0)$ for early times. The presently accepted value of the Hubble parameter is $H_0 \sim 67$(km/s)/Mpc, with 1 Mpc (Megaparsec) = 3.26×10^6 light years.

Hubble time. Estimate for the age of the universe based on the reciprocal of Hubble's parameter. At the time Lemaitre originally proposed his primeval atom theory, the Hubble time was estimated to be only about two billion years. Later revisions of Hubble's distance estimates increased the value upward to 4×10^9 years in 1948 and upward to 1×10^{10} by the time Allan Sandage took over Hubble's position at Mount Palomar.

Indeterminacy. Used to mean that Heisenberg's principle states that nature itself is "undecided", not determined, about the outcome of physical measurements of conjugated variables.

Inflationary cosmology. The cosmological theory which assumes a very early ($\sim 10^{-35}$ sec after the big bang) exponential expansion *at constant density* during which the total mass of the universe increases by many orders of magnitude. In the inflationary process the new matter comes out of the vacuum itself. The process is somewhat analogous to the continuous creation of matter out of nothing in the steady-state-cosmology of Bondi, Gold and Hoyle, which was popular in the 50's and early 60's, before the cosmic background microwave radiation was discovered by Penzias and Wilson in 1965.

Ionised atoms. Atoms under physical conditions (f.i. high radiation pressure) such that one or more electrons have been taken apart. At times below 300.000 years and temperatures higher than 3880 K the universe consisted in a *plasma* of ionised atoms, mainly hydrogen (76%) and a substantial amount of ^{4}He (23%). Later, when the temperature was cooling progressively, the universe begun to consist in a gas of neutral atoms.Substantially later, the gravitational interaction begun to pull atoms together to form protostars within protogalaxies. Finally, the universe evolved towards its present, slowly changing, overall physical conditions.

Isotropy. Attribute assumed by Lemaitre, and other cosmologists, that the universe appears the same in all directions.

To a very high degree, measurements of the cosmic background radiation confirm this.

LCDM universe. A flat universe with $k = 0, \Lambda > 0$ dominated by Cold Dark Matter.

Leptons. A class of non-strongly interacting particles which includes the electron, the muon, the tau particle, and their associates.

Magnetic monopole. An isolated pole (north or south) of a hypothetic magnet. An infinitely long magnet in which the two poles are so far apart that they do not affect each other would act as a pair of monopoles. Dirac predicted the existence of particles with properties of magnetic monopoles, and so did some grand unified theories. But they have never been observed.

Magnetic monopole problem. A problem, real or rhetorical, of traditional big bang cosmology originally pointed out by the proponents of cosmic inflation. Since the observation of real magnetic monopoles has not been confirmed, this problem appears to have faded away.

Maximum observed redshift. Largest redshift observed in light coming from very distant galaxies emitted at a time when cosmic density $(\Omega = \rho_m / \rho_c)$ was much larger than at present time. The maximum redshift observed (2014) is for the moment $(z_{obs})_{max} = 10.8$ (see Chap. 11)

Microwave. An electromagnetic wave with wavelength between one millimetre and about 30 centimetres.

Neutrino. An elementary particle of very small rest mass which is electrically neutral and very weakly interacting with other material particles. It was predicted in 1931 by Pauli and detected experimentally in 1956 by Cowan and Raines.

OFL universe. An open universe with $k < 0, \Lambda = 0$ dominated by negative (repulsive) space-time curvature (akin to radiation pressure).

Old Quantum Theory. Original Quantum Theory based upon Planck's concept of energy discreteness ("quantic character") in atoms and molecules as developed by N. Bohr and A Sommerfeld.

Open universe. A homogeneous isotropic universe in which gravity is not strong enough to stop the expansion. So an open universe goes on expanding for ever.

Phase transition. A sudden change in the properties of a material system produced by varying temperature, pressure or other physical parameter. Inflationary theory postulates a cosmic phase transition at about 10^{-37} seconds after the big bang.

Photon. A quantum (discrete) minimum of electromagnetic energy consisting essentially in a localized particle of light after the quantum theory of radiation put forward by Max Planck in 1900.

PLANCK. ESA's satellite which further confirmed in 2013 the previous results of NASA's WMAP satellite.

Planck's time. A "natural" unit of time proposed by Max Plank in terms of universal constants: $t_{Pl} \approx (\hbar G/c^5)^{1/2} \approx 5.4 \times 10^{-44}$ sec .

Quarks. Elementary constituent of protons, neutrons and other strongly interacting particles (baryons and mesons).

Red giant. A phase in the life cycle of a typical star (not very heavy, like our Sun) in which the size of the star increases tremendously and then blows up into space, leaving as a remnant a white dwarf.

Relativity. The special theory of relativity, proposed by Albert Einstein in 1905, assumes that all light propagates at a constant (invariant) speed c in free space, regardless of the relative motion of source and observer. The general theory of relativity, produced by Einstein in 1915, constitutes a general theory of gravity consistent with special relativity, and therefore, with the invariance of c. It predicts that light rays are deflected in their path when coming close to large masses. The observation of this effect taking advantage of a Solar eclipse in an expedition headed by Eddington in 1919, did make Einstein instantly famous all over the world, and confirmed the theory of relativity.

Renormalization. A theory such as quantum electrodynamics (QED) which in first approximation gives answers in agreement with experiment, in a second approximation, however, may give divergent answers (infinity) which are unrealistic. Renormalization is a mathematical transformation on the theoretical equations to avoid the unwanted infinities.

"Sensationism". Philosophical system of E. Mach which held that only sensory perceptions ("sensations") have an objective reality. Ideas or concepts, judgments connecting ideas, and logic chains of judgments have not objective reality, according to Mach.

Singularity. When the standard big bang theory is extrapolated back to time zero, a number of cosmic physical parameters become infinity, f.i. the density, the pressure, the temperature. It is physically meaningful to tray to investigate what happens as $t \to 0$ and $R \to 0$, but there is always a limit, imposed by Heisenberg uncertainty principle which sets a minimum conceivable cosmic time ($t_H \approx \hbar/M_U c^2 \approx 2.8 \times 10^{-103}$ sec) beyond which it is pointless to speculate within the realm of physical theory.

Solutions of Einstein's equation. Friedman and Lemaitre obtained rigorous solutions of Einstein's cosmological equation for a finite universe for $k > 0$, $k = 0$, $k < 0$, and $\Lambda = 0$. Using

$t_0 \cong 13.7 \pm 0.2 \times 10^9$ years and $H_0 = 67 \pm 2$ km/s/Mpc cosmic dynamics can be described relatively well with the open $(k < 0)$ solutions and the flat $(\Lambda > 0)$ solutions.

Steady state theory. A cosmological theory postulated by Fred Hoyle, Thomas Gold, and Herman Bondi in Britain in 1948, asserting that the universe is in essence without change. Hoyle, Bondi, and Gold argued against a temporal, Big Bang origin, suggesting that while the universe was expanding, the expansion occurred in a natural state similar to that suggested by de Sitter's universe, but with matter being constantly created in the form of hydrogen atoms in empty space, keeping cosmic conditions identical as a function of time. In the early 1960s the discovery of quasars and other developments in astronomy strengthened the contention that the universe in the distant past was far different from the universe at present. The discovery of the cosmic microwave background radiation all but discredited the steady state theory.

Strong interactions. The interactions which bind together quarks to form protons, neutrons, etc.

Thermal equilibrium. Physical condition which implies that photons (radiation) have a characteristic spectral distribution, the so called blackbody spectral distribution. COBE confirmed that cosmic thermal equilibrium prevailed at the time of atom formation (T~3880 K), and presumably much earlier, f.i. at nucleosynthesis, baryon formation, Plank epoch, and even before.

Type 1a supernovae. Violent stellar explosions signaling the utter demise of a star that take place when a white dwarf bleeds enough mass from a companion star to pass a certain mass limit, approximately 1.4 times the mass of the Sun (also called Chandrasekhar's limit, after the Indian-American astrophysicist Subrahmanyan Chandrasekhar), at which point it explodes. The luminance of such supernovae is so great that they can be brighter than entire galaxies as they are photographed. The consistent level of luminosity of Type la supernovae have made them an even more reliable distance indicator than Cepheids for determining extragalactic distances.

Uncertainty. Used to mean that Heisenberg's principle states that our measurements of physical observables are intrinsically limited and so that we are uncertain about the exact values of conjugated variables.

Uncertainty time. The earliest conceivable cosmic time allowed to be considered by the uncertainty principle ($t_H \approx 2.8 \times 10^{-103}$ sec).

Vienna Circle. Philosophical current grounded on E. Mach world view and originated by Philipp Franck which was the cradle of "logical positivism".

Weak interactions. Short range nuclear interactions responsible for beta decay. Neutrinos are subject only to this type of interactions.

White dwarf. Final phase in the life of a typical star after it went through the phase of red giant.

WMAP (Wilkinson Microwave Anisotropy Probe). NASA's satellite which confirmed in 2003 with greater accuracy the anisotropies in the Cosmic Background Radiation and reported for the universe an age (time elapsed since the Big Bang) given by $t_0 = 13.7 \pm 0.2$ billion years.

Ylem. The primordial cosmic material after Gamov and collaborators. When the cosmic density was incredibly high, matter density, which then could have been present in an amount comparable to that of radiation or not, was so incredibly high that far exceeded that in ordinary baryons or electrons.

Zero-point energy. Any finite amount of confined radiation in thermal equilibrium, including cosmic radiation confined by the expanding finite radius of the universe is equivalent to a set of harmonic oscillators each with quantized energy and a minimum zero-point energy. In the early *opaque plasma phase* of cosmic expansion *matter* and *radiation* expand simultaneously and total energy is conserved, and the total zero-point energy is conserved since $t_{HL} \cong 0.5 \times 10^{-102}$ s to decoupling.

After decoupling, in the *transparent phase* of cosmic expansion, as time goes on *matter* density increases with respect to *radiation* density, but, as shown in Chapter 10, the relative decrease in radiation density is complemented by a relative increase of zero-point energy density in such a way that the total energy is conserved.

AUTHOR INDEX

Adams, H., 151, 153
Alberer, M., 58
Alfonseca, M., vi, vii, 1, 111, 113, 120, 127, 128
Alfonso-Fauss, A ., vii
Alpher, R.A., 1, 2, 78, 95, 97
Anderson, C.D., 48
Aquinas, T., 4, 147
Bechart, K., 26
Bennet, Ch, 98, 111, 112, 117, 120
Bentley, R., 99, 139
Berthelot, M, 137, 138, 140
Bethe, H., 26
Bohr, N., v, 11, 13, 17, 18, 19, 20, 21, 22, 23, 24, 25, 27, 32, 36, 39, 40, 47, 50, 66, 78, 80, 81, 82, 85, 88, 130, 135, 168
Boltzmann, L., v, 8, 14, 35, 68, 69, 86, 89, 92
Bondi, H., 95, 97, 101, 166, 171
Born, M., v, 20, 32, 39, 41, 42, 43, 45, 47, 50, 51, 52, 53, 54, 66, 71, 81, 87, 90, 135, 154
Bose, S., v, 11, 64, 67, 68, 71
Bothe, W., 61, 82
Boyle, …, 139
Buridan, J., 131
Capon, L., 74
Carreira, M., vii
Chamberlain, O., 73
Clay, J., 62
Clifford, W.K., 139
Compton, A.H., v, 45, 58, 60, 61, 62, 63, 81, 82, 95, 96, 106
Copernicus, N., 9, 131
Cowan, C., 80, 81, 168
Crespo, E., 85

Cronin, …, 144
Darwin, Ch., 53, 143, 146
De Broglie, L., v, 30, 33, 81
Debye, P., v, 26, 35, 45, 52, 55, 56, 57, 58, 85
del Valle, J.C., v
Dirac, P., v, 11, 20, 24, 26, 32, 36, 40, 43, 44, 45, 46, 47, 48, 49, 50, 54, 68, 82, 129, 167
Doan, R.L., 62
Eddington, A.A., 12, 26, 81, 83, 145, 156, 170
Ehrenfest, P., 13, 71, 82
Einstein, A., v, vi, 5, 6, 9, 10, 11, 12, 14, 15, 16, 20, 24, 36, 43, 52, 53, 56, 68, 71, 82, 85, 87, 91, 92, 94, 97, 98, 100, 101, 102, 103, 104, 105, 111, 112, 113, 114, 115, 117, 124, 125, 127, 130, 131, 132, 133, 135, 140, 145, 150, 151, 156, 157, 160, 161, 163, 170
Epstein, P.S., 26
Euler, …, 99
Ewald, P.P., 26
Exner, F., 35
Fermi, E., v, 11, 26, 43, 49, 50, 63, 64, 68, 70, 71, 72, 73, 74, 76, 80
Fitch, …, 144
Franck, Ph., 13, 39, 172
Frölich, H., 26
Fues, E., 26
Galileo, G., 131
Gamow, G., 41, 50
Gárate, A., 85
Gauss, K.F., 100, 104
Geiger, H., 61, 82

Gilson, E., 137
Glashow, S., 44, 163
Gödel,, 164
Gonzalo, J. A., i, v, vi, 25, 42, 67, 83, 90, 97, 104, 110, 113, 120, 127, 128
Green, ..., 139
Guillemin, E., 26
Guth, A., 94, 97
Hagenow, C.F., 62
Hasenohrl, F., 35
Heisenberg, W., i, v, vi, vii, 3, 11, 20, 24, 26, 27, 29, 32, 38, 39, 40, 41, 42, 43, 47, 50, 52, 54, 78, 79, 80, 81, 82, 84, 91, 93, 94, 97, 103, 106, 129, 130, 131, 132, 134, 136, 154, 165, 166, 170, 172
Hilbert, D, 39
Hönl, E., 26
Hopf, L., 26
Jaki, S.L., v, vii, 5, 6, 12, 13, 14, 15, 18, 38, 46, 51, 55, 83, 85, 95, 99, 101, 104, 105, 131, 134, 136
Jammer, M., 41, 50, 54
Jaures, J., 137
Jeans, J., 50, 137
Jordan, P., v, 20, 29, 42, 43, 44, 47, 81, 154
Kant, I., 12, 99, 100, 104, 139, 141, 142
Kelvin, Lord, 139
Kepler, J., 131, 139
Kohlrausch, K.W.F., 35
Kossel, W., 26
Lambert, J.H., 99
Laplace, ..., 142

Laporte, O, 26
Lenz, W., 26
Loÿs, J.P., 101
Mach, T., v, 5, 6, 12, 13, 14, 16, 81, 170, 172
Maric, M., 9, 10
Mather, J., 95
Maxwell, C., 14, 15, 68, 69
Mendeleev, ..., 143
Millikan, R.A., 48, 62
Miranda, R., 85
Nerst, W., 10
Newton, I., 2, 8, 9, 42, 89, 92, 99, 101, 114, 131, 138, 139
Nietzsche, F., 13, 146
Norlund, M., 19
Olbers, W., 100, 101, 105, 138, 152, 163, 164
Oresme, N., 131
Pascal, B., 135
Pauli, W., v, 20, 24, 26, 39, 42, 43, 44, 52, 64, 65, 66, 67, 71, 73, 80, 129, 130, 131, 135, 168
Pierls, R., 26
Planck, M., v, vi, 5, 6, 7, 8, 9, 10, 11, 12, 13, 14, 15, 16, 27, 36, 40, 48, 52, 53, 56, 57, 68, 76, 78, 82, 84, 85, 86, 87, 88, 89, 90, 91, 92, 93, 94, 96, 97, 112, 113, 130, 131, 132, 133, 135, 155, 159, 164, 168
Popper, R., 152
Raman, C.V., 53
Reines, F., 80, 81
Riemann, B., 100, 139
Riess, A.G., 112, 117, 120, 125, 128

Rogowski, W., 26
Rubens, H., 8, 86
Rutherford, E., 19, 20, 24
Salam, A., 44, 163
Sandage, ..., 112, 166
Schwarzschild, K., 51, 95, 96, 115, 116, 133
Schwinger, J., 3
Seeliger, ..., 26, 139
Seeliger, R., 26, 139
Segre, E., 73
Simon, A.W., 61
Smoot, G., 95
Sommerfeld, A., v, 17, 25, 26, 27, 28, 39, 55, 66, 88, 154, 168
Sorel, G., 100
Spencer, H., 15, 142, 143, 144

Spergel, D., 98
Stark, J., 10
Stuart Mill, J., 151
Thompson, J.J., 19
Tinsley, B., 112, 120
Voltaire,, 99, 138
Weinberg, S., 44, 163
Welker, H., 26
Wentzel, G., 26
Whitehead, A.N., 147
Wien, M., 35, 90
Wigner, E.P., v, 10, 42, 44, 50, 63, 64, 75, 76
Wigner, M., v, 10, 42, 44, 50, 63, 64, 75, 76
Wilson, C.T.R., 61, 63, 95, 97, 112, 157, 166

SUBJECT INDEX

Aachen, 26, 55
Accelerated expansion, 127
Acheson-Lilienthal plan, 23
Antimatter, 49
Artificial radioactivity, 73
Ateneum, 15
Athens, 33, 153
Baruch proposals, 23
Berlin, 7, 9, 10, 36, 40, 52, 57, 76, 155
Big Bang, 1–3, 92, 95, 97, 102, 107, 114, 157, 159, 164, 168, 171
Blackbody, 170
Bose-Einstein condensation, 69
Bose-Einstein statistics, 11, 71
Breslau, 35, 51, 52
Brownian motion, 10
Brussels, 33, 88, 129
Bucharest, 33
Budapest, 75
c (velocity of light), 8
Cambridge, 19, 47, 50, 52, 53, 61, 97, 151, 156
Cavendish Laboratory, 19
CERN, 23, 155
Chance, 53, 153
Chemistry, 26, 59, 85, 132
Chicago, 5, 13, 15, 52, 61, 72, 76, 83, 136, 152, 153
Classical Mechanics, 132
Clausius, 7
Clausthat, 26
COBE, 96, 102, 159, 162, 164, 170
Columbia University, 72
Compton and Schwarzschild radii, 96
Compton effect, 61

Copenhagen, 11, 18, 19, 20, 39, 40, 66, 78, 146, 153
Copenhagen School, 20
Cornell University, 58
Cosmic Background Radiation, 95, 102, 157
Cosmic density parameter, 160
Cosmic zero-point energy, 106, 110
Cosmological constant, 161
Dark energy, 121, 124, 161
Dark matter, 113, 167
Drude theory, 26
Dublin, 36
Eiffel Tower, 31
Electromagnetism, 132
Electrons, 62
Energy conservation, 163
Euclidean infinity, 138, 143, 150
Fermat Last Theorem, 130
Fermi-Dirac statistics, 11, 49
Finite universe, 133
First World War, 35
Flat universe, 119, 164
Florence, 46, 71
Foucault currents, 56
Fourth Lateran Council, 148
French Academy, 30
General Relativity, 12, 100, 104, 140, 161, 164
German uranium project, 40
Ghent, 36
Gospel, 14
Gottingen, 6, 26, 39, 47, 51, 62, 53, 54, 56, 66, 71, 77, 85, 104, 154, 155
Graz, 36
Hamburg, 43, 66

Heisenberg-Lemaitre units, 93
Helgoland, 39
Hiroshima, 10
Implications of a finite universe, 98
Inflationary cosmology, 166
Institute of Advanced Studies, 10
James Webb Space Telescope, 112
Joule Memorial Lecture, 101
Kaiser Wilhelm Institute for Physics, 40
Keplerian velocities, 121
Kiel, 7
Lambda Cold Dark Matter universe, 113, 134
Lausanne, 33
LCDM model, 113
Legion d'Honneur, 33
Leipzig, 40, 57, 73, 90, 100, 139, 154
Leyden, 71
London, 67, 83, 110, 139, 157
Los Alamos, 23
Maastricht, 55
Madrid, 78, 83, 85, 90, 97, 98, 110
Manhattan Project, 23, 72, 76
mass of Neptune, 14
Max Planck Institute, 40, 57, 155
Medieval Christian Europe, 13
Michigan, 66, 74, 104, 136
Milky Way, 104, 132, 139, 159, 162
Millikan oil-drop, 62
Modern Physics, 5, 15, 132
Mount Wilson Telescope, 112

Multiverse, 151
Munich, 7, 9, 26, 38, 39, 41, 56, 66, 139, 154, 155
Nagasaki, 10, 63
NASA, 98, 112, 117, 168, 171
Neutrino, 95, 171
OFL model, 113
Old Quantum Theory, 27, 168
Open Friedman-Lemaitre universe, 111
Open universe, 117, 168
Optics, 132
Oxford, 36, 53, 90, 110, 151
Paris, 30, 31, 33, 151
Pasadena, 48
Periodic Table, 72
Physics and Philosophy, 18, 41, 50, 129
Pisa, 71
Planck Satellite, 112
Prague, 10, 13
Primeval Atom Theory, 95
Princeton, 9, 10, 16, 36, 61, 66, 76, 128, 135
Principle of Complementarity, 21
Purdue, 66
Quantum Electrodynamics, 42
Quantum Mechanics, 3, 8, 11, 17, 20, 27, 29, 40, 41, 42, 43, 45, 47, 50, 53, 64, 78, 130, 155
Quebec, 33
Rockefeller grant, 39
Rome, 162, 164
Royal Danish Academy of Science, 23
Schwarzschild radius, 95, 115, 116, 133

Schwinger effect, 3
Science and Religion, 131
Scientific Relativism, 15
Second Principle of
  Thermodynamics, 7
Second World War, 6, 10, 14
Seton Hall University, 85
Singularity, 169
Solar System, 142, 147
Solvay Conference, 88, 129
Sorbonne, 31
Special Relativity, 10
St. Louis, 61
Steady State Theory, 96, 101,
  157, 170
Stuttgart, 35
Swedish Academy of Sciences,
  32
Swiss Federal Technical
  Institute, 56

Theory of Relativity, 5, 12, 66,
  105, 169
Thermodynamics, 13, 132
UAM, 85
UNESCO, 32
United Nations, 23
Utrecht, 56
Vatican I, 148
Vienna, 13, 34, 35, 37, 65, 171
Warsaw, 33
Westinghouse Lamp Company,
  61
WMAP, 98, 102, 112, 113, 168,
  171
Wooster, Ohio, 60
X-ray research, 57
X-rays, 57, 61, 62, 82, 137, 162
Zero-point energy, 172
Zionism, 10
Zurich, 35, 51, 56, 57, 66, 6

Printed in the United States
By Bookmasters